LANDSCAPE, RACE AND MEMORY

Heritage, Culture and Identity

Series Editor: Brian Graham,
School of Environmental Sciences, University of Ulster, UK

Other titles in this series

Geographies of Australian Heritages
Loving a Sunburnt Country?
Edited by Roy Jones and Brian J. Shaw
ISBN 978 0 7546 4858 1

Living Ruins, Value Conflicts
Argyro Loukaki
ISBN 978 0 7546 7228 9

Geography and Genealogy
Locating Personal Pasts
Edited by Dallen J. Timothy and Jeanne Kay Guelke
ISBN 978 0 7546 7012 4

Jewish Topographies
Visions of Space, Traditions of Place
Edited by Julia Brauch, Anna Lipphardt and Alexandra Nocke
ISBN 978 0 7546 7118 3

Southeast Asian Culture and Heritage in a Globalising World
Diverging Identities in a Dynamic Region
Edited by Rahil Ismail, Brian Shaw and Ooi Giok Ling
ISBN 978 0 7546 7261 6

Valuing Historic Environments
Edited by Lisanne Gibson and John Pendlebury
ISBN 978 0 7546 7424 5

Sport, Leisure and Culture in the Postmodern City
Edited by Peter Bramham and Stephen Wagg
ISBN 978 0 7546 7274 6

Culture, Heritage and Representation
Perspectives on Visuality and the Past
Edited by Emma Waterton and Steve Watson
ISBN 978 0 7546 7598 3

Landscape, Race and Memory
Material Ecologies of Citizenship

DIVYA PRAFUL TOLIA-KELLY
University of Durham, UK

Routledge
Taylor & Francis Group

LONDON AND NEW YORK

First published 2010 by Ashgate Publishing

2 Park Square, Milton Park, Abingdon, Oxon OX14 4RN
711 Third Avenue, New York, NY 10017, USA

Routledge is an imprint of the Taylor & Francis Group, an informa business

First issued in paperback 2016

British Library Cataloguing in Publication Data
Tolia-Kelly, Divya Praful.
  Landscape, race and memory : material ecologies of
  citizenship. -- (Heritage, culture and identity)
  1. Asians--Great Britain--Ethnic identity. 2. Women
  immigrants--Great Britain--Psychology. 3. Memory--Social
  aspects--Great Britain--History.
  I. Title II. Series
  305.8'914041-dc22

Library of Congress Cataloging-in-Publication Data
Tolia-Kelly, Divya Praful.
  Landscape, race and memory : material ecologies of citzenship / by Divya Praful Tolia-Kelly.
    p. cm. -- (Heritage, culture and identity)
  Includes bibliographical references and index.
  ISBN 978-0-7546-4957-1 (hardback) -- ISBN 978-0-7546-8823-5 (ebook) 1. Human geography--Great Britain. 2. Human geography--India. 3. Landscape assessment--Great Britain. 4. Landscape assessment--India. 5. Geographical perception--Great Britain. 6. Geographical perception--India. 7. Women immigrants--Great Britain--Psychology. 8. Women immigrants--Great Britain--Social conditions. I. Title.

  GF551.T65 2010
  304.20941--dc22

                                                              2010006001

ISBN 13: 978-0-7546-4957-1 (hbk)
ISBN 13: 978-1-138-25496-1 (pbk)

Contents

List of Figures

Appendices

The author has made every effort to recover and credit copyright holders. Anyone objecting to the reproduction of any copyrighted images without permission should contact the author.

Acknowledgements

Thanks to Jacquie Burgess for her love and concern always.

A special thank you goes to my editor Val Rose and my 'book buddy' Alex Hall. Several big special metaphorical hugs go to Julia Crew, Mike Crang and James Kneale, all of whom have listened, kept me laughing, smiling, and believing in myself.

The really special sociology crew includes three stellar women who got me through the tough times of academic isolation and fear: Mimi Sheller, Anne-Marie Fortier and Sara Ahmed have been wonderful and principled and have given me courage to keep going as a black woman in academia, despite the terrors. Also along the way Gail Lewis, Sonya Sharma, Vanessa Castan-Broto, Iain Chambers and Lidia Curti have guided and enriched my intellectual journey.

I would like to acknowledge the support and care of the late Professor Denis Cosgrove. I met dear Denis the week that my father died, and for years afterward he gave me support and empathy, and continued to provide guidance until his own death in March 2008. Denis taught me the value of politics and integrity and also to engage with what matters. The landscape politics, cultural geopolitics and historical ecologies of the visual embedded here in this research are inspired by his teachings. I hope they do him justice. Thanks, Denis, for your belief in me and encouragement.

I thank my friends and academic colleagues who have variously offered their time and support over the years without always realizing it; these include: Michael Keith, Gillian Rose, Jane M. Jacobs, Christina De Matteis, Steve Pile, Julian Agyeman, Eric Laurier, Susan Mains, Phil Crang, Gareth Hoskins, Kathy Burrell, Antonia Noufsia, Stuart Elden, Ash Amin, Susan J. Smith, Alastair Bonnet, Anoop Nayak, Peter Hopkins, Nick Mann, Deborah Thien, Ian Cook, Andy Morris, Tim Cresswell, Pete Merriman, Uma Kothari, Adam Holden, Kate Holden, Richard Kensington, Simon Bowhill, William Cronon, Tim Edensor, Alison Blunt, Miles Ogborn, Catherine Nash, Dipti Bhagat, Stephen Daniels, David Matless, Steve Legg, Alex Vasudevan, Luisa Bailaweics, Richard Munton, Peter Wood, Robyn Longhurst, David Pinder, Melanie Carvalho, Graham Lowe, Hamzah Muzaini, Kath Ray, Sharon Cowling, Gavin Brown, Jason Lim, Kath Browne, Alison Williams, Kathrin Horschelmann, James Glendinning, Amanda Bingley, Sara Kindon, Nicole Cook, Cindy Webber, Mags Adams, David Gilbert, Emily Gilbert, Dave Pomfret, Alex Clarke, Sarah Atkinson, Derek Gregory, Nirmal Puwar, Mark Boyle, Mary Gilmartin, Patricia Noxolo, Parvati Raghuram, Robina Mohammad, Yok Sum Wong, John Urry, Adam Tickell, Roger Lee, Richard Dennis, Aoibheann Kilfeather, Joseph Garver, Francisco Klauser, Julian Fisher, Rob Kitchin and Jayesh Ghelani.

I dedicate this book to my family, most importantly my mother Pratibha Tolia, without whose daily care and tenderness I could not be me, and my late father Praful Dalichand Tolia, without whose inspiration I could not research.

Nileshvari and Manisha, thanks for sharing the joys and struggles of daily life.

And many thanks, infinite kisses and much love to Pete, my real-life soulmate.

Chapter 1

Ecologies of Citizenship:
Landscape, Race and Memory

Natural Citizenship

On 12 October 1979, my father, Praful Dalichand Tolia, was sent a certificate of naturalization for himself, my mother Pratibha and my then only sister. We had been rubber-stamped. It was official. We were accepted and considered naturally British. When I think of this now, it seems that I had officially passed the tests. I'd been through a bureaucratic catalyst, and had changed from a non-natural, *non-native* to British soil (namely a 'foreign' citizen of the British East Africa Protectorate) – and my parents too – to a bona fide citizen of the United Kingdom and its territories. My father deemed this a formality, as we were all born and brought up as *feeling* British; in Kenya, we had always felt protected by the wealth of infrastructure and governance systems of Elizabeth II, Queen of England. I had been born in Nairobi, Kenya, and my parents in Mantarre, Uganda, and Dar Es Salaam, Tanzania, respectively. We were all living near the Equator, in the Northern Hemisphere, yet were acclimatized to speaking English, drinking tea, and learning about English manners and the cultural praxis of Englishness through reciting English nursery rhymes, eating greens, recognizing gingham, reading C.S. Lewis and Rudyard Kipling, eating sandwiches, enjoying cricket and having picnics at Lake Nevasha. Although some of these things are done the world over, for my family, these customs were our *Englishness.* The naturalization certificate validated our being proper subjects of the Queen, after years of subjugation to English law, governance and systems of social mobility in the British East Africa Protectorate. My family and I have always been grateful to the Crown; my mother says that if our expulsion had occurred today, she wouldn't be confident that the avenues of attaining a place to call 'home' in the UK would have been offered. We would have been processed as *asylum seekers*, facing the precarious existence that this immigration status encompasses.

Our family's biological and cultural roots were in Indian soil, several generations prior. Our blood was linked to Indian Rajputti families of the eighteenth century, and had flowed through practitioners of both Jain and Hindu religion when deemed politic and even sometimes as a means to avoid genocide. Our last territory of citizenship had been Kenya; we loved the air, earth, flora and fauna of the African savannah and enjoyed the spectacle of the Masai, wild safaris, the Rift Valley and the mountainscapes of Kilimanjaro and Mount Kenya. At once these were our cultural attachments to landscapes adopted as our own, simultaneously these locales were figured through textual and musical attachments to the Beatles, Shakespeare, Dickens, *Alice in Wonderland*, Constable, Wordsworth, Johnny Mathis, Kenny Rogers, Chubby Checker and Elvis. These cultural texts evoked landscapes, ecologies and peoples that were 'Other', but part of our quotidian scene. Our cultural citizenship was not contained within a single territorial culture. Geographically, British East Africa had incorporated present-day nations such as Uganda, Kenya and Tanzania. These were imagined as one single national community. My father had travelled from Mwanza to Mantarre and ended up in Nairobi. These three cities being in three separate nations was not a factor of concern. How could he then situate a singular notion of national citizenship in the twentieth century? Connections to land, blood and soil across India and British East Africa were core to his being, his material body, and his imagined community of fellow citizens. His connections with citizenship were *ecological*, materially, textually and conceptually. The state boundaries of India and British East Africa had changed several times in his lifetime. The immigration and citizenship legislation of Britain, too, had changed. He had been at various moments a different version of 'British'. He had been expelled from Kenya in 1972, along with my mother and their two daughters, for not being Kenyan. They at once were neither 'Kenyan' nor 'British', and they waited from March 1972 until 1979 to be secure in their status. Since this expulsion, and the correspondingly cool welcome in Britain, we have never forgotten that citizenship is ephemeral. As racialized bodies, we will always be 'Other' to the core values of 'Indianness', 'Englishness' and 'Africanness'. In this respect, we have found solace in the fact that we did survive and that we have benefited from the varied life experience and resulting resilience that is part of being British Asian. This book is about the process of becoming a citizen; of becoming naturally *in place*. It is also engaged with the ways in which citizenship can only be understood ecologically, materially (as bodies in the world), texturally (through cultural texts and textures of inhabitation), and conceptually (as a multidimensional way of thinking being, living and feeling). My family's journey represents the journey for many; through colonial territories, various biomes, and the environmental, spiritual, emotional and economic adjustment to that mobility. My family has been part of a racialized minority in Britain which were once a racialized minority on the African continent, but were also part of a majority; they formed the bulk of those British citizens who did not live here – they were those living in the colonies, populations of *British* who had not known or experienced this land, but had contributed to its wealth and success through their labour, embodied attachments, innovation, taxes, motivation and hardiness. These populations continue to be both *inside* and *outside*. Some of those who are British Muslims have been, since 9/11, more *outside* than others. Yet they are here because our Sovereign ruled *over there* with much violence, oppression and subjugation, the legacies of which continue to haunt us, creating the unstable and very *live* politics of citizenship that is practised on the soil, within the culture, and through infrastructures of economy, politics and heritage.

The process of *naturalization* is equally applied to the non-human world. The Royal Horticultural Society holds a definitive list of those plant species territorialized on British soil that are officially *natural* in Britain, but that were not always *native* to Britain. The general belief is that non-native species are a threat and that native species are under threat from aggressive, organic 'miscegenation' and thus the ecosystem of our native land needs to be preserved from the threat of 'foreign' invasion. However within these narratives are notions of plants that were non-native being *naturalized.*

Invasive non-native species are considered to be the second most significant cause of biodiversity loss on a global scale. The threat from non-native species is increasing with the continued growth of global trade and travel (Natural England, 2006).

An example of this is the walnut tree, which, after being called a 'foreign species', is now regarded as natural to the UK landscape. According to the Royal Horticultural Society, the walnut is an example of tree species that although described as 'exotic', has stayed so long that it is now part of our understanding of what constitutes as 'native landscape' (see Tolia-Kelly, 2007a). The notion of a 'natural' ecology or 'natural' citizen in both human and non-human cases rests on cultural and political definitions. An ecological attitude to both attests to the need for our embracing of a tolerant attitude to both, where a natural equilibrium will prevail. Or indeed, we need to recognize that terms such as 'native' and 'non-native' are culturally defined.

> We are a genus of one species with many subdivisions, and there are divisions of that species that have predispositions that could prove fatal to the other subdivisions of that species. In the big picture, the exposure of a species to all there is to be exposed to ultimately will give it greater strength. ... As an individual you stand on your own two feet and are confronted with the world. Any claim that you shouldn't be there is just wrong. You might not like the fact that there are gay people or those who have a different skin colour – these are elements of the human condition. Being exposed to diversity can only help you, strengthen you – exposure or collective living can retain integrity for the species – I would argue that this is the same for any other organism. (Interview with Phil Dewhurst, arboriculturist, Burnley, UK, July 2004; see also Tolia-Kelly, 2007a)

What Phil Dewhurst's statement makes clear is that by embracing an 'organic cosmopolitanism', new notions of ecological citizenship are possible, especially one that can bridge the mobility of human and non-human species. This would be called an ecological citizenship based on a natural cosmopolitanism (see Clarke, 2002; Turner, 2002).

Doing Ecologies

The aim of this monograph is to bring together ideas and research on race, ethnicity and cultural geographies of citizenship and identity. Citizenship here is ecological: connections with soil, landscape, and the iconographies of lived experience are central. Our identification with these ecological textures is the basis for identity, belonging and embodied connections to place, space and nation. The monograph offers ways to think about race and citizenship in the twenty-first century through a commitment to culture, materialism and a notion of environmental histories, and thus it is framed as thinking about identity ecologically. Doing ecologies of citizenship and identity is about embracing a multinodal approach to thinking about the subjectivities of the racialized body in the 'West'. The 'West' is in parenthesis here because the idea of the West is fissured with people, ecologies, knowledges, value systems and materialities of that which is constructed as 'East', and thus cannot attain integrity or currency without these presences. The idea or political notion of there being an inextricable synthesis between these two co-constituted sites, natures, cultures and societies is illustrated throughout this research monograph; the embodied identities of those positioned as 'Other' are created and are part of colonial, imperial and academic imaginaries that have dominated the last few centuries. The Occidental, or that which is English, is reliant upon, and is structured through, the 'Oriental', or 'Indian' or 'African'; these material and cultural connections are running through its veins, keeping it alive and well.

The research presented here is based on 10 years of working with South Asian women living in London. All of these women are first-generation migrants, some of them twice or triple migrants. All have lived under the auspices of the British protectorate in East Africa, colonial rule in India (before partition), and colonies in Yemen, Zimbabwe, Malawi and many others. These women are at the heart of Englishness, as made through its legacy in Empire and the landscapes of Empire. Their biographies reflect the footprint of the British Empire, its economies, natures and cultural sphere of influence. Their migration reflects the Empire's need for the movement of labour, the use of colonized citizens to develop economies, to expand Empire, to consolidate geopolitical ambitions, and to instil acceptance of a model of law, economy and a political identity of Britishness itself. This research is about thinking Britishness within the boundaries of British citizenship, to unravel the multinational nature of Britishness and British culture. The identities of these South Asian women are shown through their very connectedness with landscapes, itself a European invention exported internationally. The iconographies of landscape, through which these individuals articulate their political, cultural, religious and social citizenship, are made tangible through the use of visual methodologies, material methodologies and in-depth group workshops on biographical routes of migration to the UK. The aim was to think 'identity' through geographies of landscape, nature and coordinates of everyday lived experience. This is counter to a body-centred approach to identity, reifying the notion of 'South Asian', 'Afro-Caribbean', 'African' and others. The body in a situated experience of landscape is what is pursued through ethnography.

Diaspora Identity

As I write this chapter, 'diaspora' research is on an exponential rate of increase in terms of both publication and funding priorities. Its use and meanings have become diverse (Brubaker, 2005). My research has been fully shaped and driven by 'diaspora' as conceptualized by Stuart Hall (1990), William Safran (1991), Paul Gilroy (1993a) and Avtar Brah (1996), none of whom had worked with notions of a visual and material citizenship defined through encounters with landscape. Diaspora is a term that is both helpful and reductive. In terms of visualizing a society of connected, mobile migrants or exiles, it instantly enables the possibility for a theorization of a particular and specific culture of experience, memory, political positioning and imaginative geography. What 'diaspora' often conceals when used in the contemporary social sciences is the violence of racial oppression, alienation and 'positioning' that is present in earlier accounts by Hall (1990), Morrison (1990), Kitaj (1989), Araeen (1992a, 1992b) and Hobsbawm (2005). Diaspora as a contemporary concept conceals the processes of racialization not intentionally but through slippages that occur when thinking about 'diaspora migration' 'culture' and 'landscapes'. The intention behind the title of the book, *Landscape, Race and Memory*, is to reassert the critical process of racialization that enforces upon the diaspora a positioning of always needing to respond as 'Other', as in Said's (1978) thesis. The diasporic sensibility is a sensibility that develops from experience of a racialization, and a mobility combined. Embedded in the diaspora sensibility are experiences of fear, insecurity, negation, abjection and in-betweenness in varying proportions. There is a biological aspect of diaspora racialization that is inherited from race taxonomies of the colonial period. These, like the diaspora, shift in formation, response and integrity. In response to this racialization, the diaspora develops a sense of 'imagined community' (Anderson, 1991), 'belonging', and what I posit in this research as *citizenship* that is *environmental* and *ecological* as a response to being obscured and left at the social, political and cultural margins of 'others'. In this case, then, my consolidation of an *Asianness* is not to create a fantastic, coherent notion of 'diaspora' culture, but to affirm a notion of diaspora as continually in violent circumstances, which figures diaspora citizenship in dynamic flux, and defined 'in-relation' to an *Englishness* (Matless, 1998) which is also in tension with the notion of an inclusive Britishness. This conceptualization of diaspora does, however, resonate with contemporary writings on cosmopolitanism and transnationalism (see Turner, 2002, 2006). In this research the cultural and geographical diversity within the British Asian diaspora is made tangible, yet the integrity of diaspora community is also expressed. Environmental citizenship in this research is a conceptual understanding of the diaspora's coordinates of living and being a citizen within an environment. There has been rupture of identification and experience between past landscapes and present ones; soil, territory and a sense of 'home' are at stake. Therefore, what occurs in this research is a mapping of a territory of citizenry that operates beyond a formal timeline of history, heritage, cartography and spatial understandings of the 'body politic'. What you are drawn into is a time-space set of coordinates that, through visual, material and biographical texts, enables an engagement with the landscapes of citizenship that enable diasporic expressive cultures of Britishness and citizenship.

Research on identity cannot be separated neatly into research on the body without embedding within it a notion of memory, or of culture without considering economy, or any tropes of identity without thinking about locales of belonging. Body, spirit, intellect, the affective and economies of being are linked through 'structures of feeling' (Williams, 1977), being, remembering, dwelling and landscapes of living. Doing ecologies of race is recognizing that questions of identity are riddled with the false logic of race, but that these are live and conjoined in contemporary politics of national citizenship, of global mobility and the political and moral geographies of belonging, being and feeling. Here, it is important to move away from the reification of categories of 'Black' and 'White' without rejecting the reasons for their resonance in contemporary society. It is also important not to relapse into the particular type of divisive identity politics of the 1980s, which served to bolster notions of 'type', racial or cultural 'competences', 'identity boundaries' and biological identification practices. The complex of identity is connected with global histories of the movement of land masses, genes, peoples, their regimes of cultural signification, and the political strategies for resource management and control. Ecological thinking allows us to embrace these complexities, which honour the histories and myriad differences that exist between human groups, but also to be genuine about the histories, materialities and dynamism of the categories of ethnicities, peoples and 'races'. Situating these notions of 'identity' can only be worked through using geographical and temporal frameworks; phenotype, environment and history are rejected for the uses of the intellectual bodies of knowledge on landscape, race and memory.

Ecological thinking positions the 'West' within a universal plane of thinking about culture, identity, landscape, and language. This thoughtful approach retains an imperative to acknowledge power differentials, economic unevenness, lack of global distributive justice, and indeed uneven access to a societal voice or power. By thinking of identity as ecologically positioned, we can be reflexive epistemologically, ontologically and heuristically. The avoidance of epistemic violence (see Code, 2006) is paramount. So a participatory approach to ethnography, storytelling, historical writing and academic dissemination moves a little way towards eliminating some of the literary and material violences which occur against the 'East', the 'Other', 'Muslim', 'famine victim', 'asylum seeker', 'migrant' and other outsiders. Ecological thinking is not a didactic mode of enquiry; instead, it is philosophical advice. Some would call this an ethics of practice, and I would call it responsible political practice in academic research.

The methodologies employed in this research are intended to be complicit with an ecological philosophy of thinking, both necessary and endemic to an ecological approach to identity. The visual and material elements of our lived daily lives are integral to our political, biological and sensory matrices of being and negotiation. There are accumulated knowledges that are corporeal, habituated and felt, embedded in the rhythms, spaces and flows of our lived identities. The visual, material, temporal, geographical and sensory nodes of our identities are critical in a mapping of identity through power, body memory, biological identity and sense of place in the world. These nodes form a continually evolving and dynamic matrix of being that effectively binds us to opportunities, capacities and mobilities in the modern world. This research privileges the nodes of landscape, race and memory to attempt to refract and triangulate the mobilities, materialities and lived encounters of racialized populations of post-colonial citizens of Britain.

The importance of the 'textural' is that, as embodied citizens, we engage holistically with our locales of being and living. Therefore, thinking identity purely through the skin, the visual or sound, for example, would necessitate a reductive approach to cultural identification and thus also to the politics of representation and race. The need for a more triangulated approach to identity is promoted here, through coordinates of *living*, and through the visual and material cultures of the home as being cultural artefacts of diasporic heritage, thus being critical to a domestic landscape of belonging. The use or 'lens' of the 'textural' offers a dimension of an embodied approach to identity, where identity is not delimited to skin, nationality, economic status (as in the 'Third World'), political citizenship or religion. A notion of diaspora identity in the case of the South Asian women in this research enables us to think about their 'positioning' (Hall, 1990) in respect of lived biographies and their relationship to national identity through cultural artefacts, landscape paintings, oral histories and sensory memories. Thinking ecologically enables us to situate the racialized identity within a framework of the temporal, spatial, and sensorial, and the self-imagined or self-determined realm of identity praxis. The materials specifically engaged with here are visual, artefactual and imagined landscape connections which are biographical, utopian and lived.

Both For and Against 'Black Minority Ethnic'

Throughout this monograph I will use terms such as 'Black', 'race' 'ethnicity' and 'identity'. For many contemporary post-structuralist thinkers, these terms are no longer 'live' or indeed valuable. In my theoretical argument, thinking ecologically requires the reader to engage with power geometries (Massey, 1993), the continuing use of 'race' as a structural strategy of exclusion, and the oppression that is present in the everyday lives of racialized communities. 'Race' has been dismissed as being an inaccurate and non-scientific category; namely, culturally constructed and reproduced. Historically, 'race' was made a scientifically driven account of human hierarchy, of both intellectual and physical capacity. In the latter part of the twentieth century, after 1945, much of sociology and cultural studies has sought to challenge these categories under which colonial and imperial taxonomies of race, national type and environmentally determined ethnicities were framed. More recently, these challenges have been translated to dialogues within anti-racist groups and policymakers, and a challenge to the categories and reifications of non-White groups and ethnicities developed during the realm of multiculturalism and the anti-racist movements from the 1970s onwards. There has been a struggle against non-positivist accounts of ethnicities that are plotted, mapped, defined and profiled under the auspices of 'migration research' (Modood, 1998). Alternatives to the framings of race and ethnic studies have emerged both from the radical activist's perspective and through post-structuralist critiques and post-colonial theory. In post-colonial accounts of race and difference, culture is the point of examination, namely literature and the voice of the subaltern. This is opposed to the nineteenth-century science of taxonomies, categories, environmental determinism and miscegenation (Anderson, 2007). In the twentieth century, namely the 1970s, the focus of anti-racists was also on the body, skin colour and notions of universal origins in African territory. For some, these new ways of thinking race were liberating and held real political currency. However, in light of the cultural studies school of thinking (Stuart Hall, Paul Gilroy, Iain Chambers, Lidia Curti, etc.), these anti-racist lobbying groups' categorizations were also reductive and intellectually pedestrian. Despite being rooted in social justice, and claims for the redistribution of wealth, social infrastructure and rights, the language of identity politics did not resonate for all of the oppressed. Political struggles that centred on single-issue politics around skin colour, gender, disability, and sexual orientation often excluded questions of geography, class, history, and self-determined identity politics. All of these movements retained notions of identity as singular bounded categories Part of the problem here was that 'identity' was crude, clumsy and outmoded. This sense of the unfair homogenization of groups' identities was damaging and disempowering, resulting in fractious and troubled times for anti-racists. This period of time coincided with the emergence of a body of thought labelled 'post-colonial theory', which recognized one levelling means of oppression above all others, that of colonial rule and its subjugating mechanisms of disempowerment. Post-colonial bodies had been subjugated, oppressed and silenced in particular ways, including the very real crushing of language systems and systems of cultural expression, heritage and dynamic self-determination. For writers such as Guyatri Chakravorty Spivak (1988), not only had there been suffering, but also the material history, the cultural, biological and intellectual integrity of a colonized society, had been smashed. This theory took the movement of anti-racism into the realms of acknowledging cultural, ethnic and intellectual violence alongside the physical violence of the street fascists in Europe and the US. What is important here is that post-colonialism was liberating. It allowed intellectual discussions about the 'hybrid', the voiceless 'subaltern', and the nature and mechanisms of imperial subjugation. This paralleled the recognition that racism was also evolving into the realms of *cultural difference* from being solely about *biological difference* (Gilroy, 1993b). The theory of a post-colonial scholarship was enabling and visionary. However, it was still locked into the elite world of literary theory and citations (Said, 1983, p. 2). The chasm between anti-racists on the ground and the post-colonial thinkers has not always been bridged. The aim of thinking *ecologically*, therefore, is precisely about situating the post-colonial subject in the quotidian landscape. The aim is to think about the subject as figured through biography, language, heritage, memory, culture and nature. This enables us to consider a post-colonial *positioning* through the lens of power, geography and time. *Ecological* thinking considers structure, language, economy, mobility, power, race and relationships *in situ*, within landscape politics.

Thinking Politically About Doing Theory

One important issue in this monograph is that thinking about identity, race and belonging is theorized and worked through by using situated methodologies and through praxis. The promotion of ecological thinking here is worked through three modes of *doing theory*, using three modes of methodological practice. These methodologies allow space for a synthesis between theory and method to become clear. Also this offers a route of accountability and exposure of the *messiness* of the process in the field. The advocating of theoretically grounded practice aims to challenge universalisms in philosophies and theories. Occlusions of matrices of power within theoretical constructs result in universalist and ethnocentric theorizations. As an example of the manifestation of these universalisms, the dangers of universalism are mirrored in the political rhetoric of neo-conservatism, post-9/11. In this period we have endured metonymical slippages in provocative pronouncements of what 'others' and 'terrorists' are. These slippages are where a universal figure of 'non-patriot', 'bomber' and 'Muslim cleric' supports a climate of fear and loathing of any number of bodies that do not slip back into being figures of acceptable, lovable citizen (Ahmed, 2004a, 2004b). When this figure is universally identified, this body is disorientated and unmappable; 'it becomes difficult to locate, situate, personify and identify' (Weber, cited in Ahmed, 2004a, p. 135). This lack of specificity denies the fact of our myriad abilities to move – and be feared, loved and hated – within the social sphere into a world where universal types are operative and the legacy of cultural theory has been lost.

To counter these universalist imperatives, the power of spatial politics is pivotal to the materializing of particular geographies that reduce material encounters to categories of 'race', 'gender' and 'sexuality' (Saad and Carter, 2005). Writing on 'whiteness' (Bonnett, 1996, 1997, 2000) has also been central to thinking through and challenging assumed racial universalisms that exist in visually dialogical media that shape our world (Dyer, 1997; Neal, 2002) It is thus critical to think plurally about theory. One set of writings that have argued this has been that on race and racisms (Gilroy, 1987, 1993a, 1993b; Goldberg, 1993; Hall, 1990, 1996, 1997; Hall and Du Gay, 1996; Solomos, 1993; Solomos and Back, 1995). This monograph seeks to take these ideas further through drawing on the connections between nodes of 'landscape', 'race' and 'memory' as experienced by racialized, post-colonial migrants, living in the everyday landscapes of north-west London.

Memory and Landscape

Memory in this monograph is a means to examine the ways in which post-colonial citizenship in Britain is experienced – through remembered citizenships of 'other' geographies abroad. In this book I reflect on how cultural landscapes of British Asian women, as a mode of enquiry, can reveal cultures of memory and social-historical narratives about migration, citizenship and belonging. Landscape research and writing within geography often focus on cultural representations that are not normally inclusive of everyday, valued landscapes of mobile, racialized migrant communities. The cultures of the British landscape are embedded with ideas of heritage, national identity and citizenship and often centred on 'English' cultures rather than 'British' cultural citizenship. This book seeks to explore what David Matless (1998) termed 'Englishness in variation' from a post-colonial perspective. British Asian women hold embodied relationships with cultures of 'landscape'; however, these are rarely encountered. By engaging with them closely we can consider them as counter-narratives to popular conceptions of landscape as a cultural 'field of vision'. The research with British Asian women aims to advance research on landscape and memory using a post-colonial view, particularly through the doubly marginalized voice of Asian women. This focus on the value of landscape is considered here as a means of narrating social history, and is termed *memory-history*, thus locating a cultural citizenship in geographies beyond those traditionally encountered as 'English' in formal accounts. Through memory narratives, *Englishness* itself is depicted as formed through a circulatory route that has traversed the colonial pathways of commerce, labour and political rule.

Post-colonial *Englishness* is presented in the oral testimonies, material cultures and visual materials produced in this research. The research grounds the women's landscape connections and memories and argues that these reflect an Englishness that is post-colonial – both politically and geographically (see Blunt, 2003; Blunt and McEwan, 2002). In particular what this research has achieved is to attend to the diasporic migrants' experiences highlighted by Alexander (1996), Bhachu (1985), Brah (1996), Kalra et al. (2005), Modood (1992), Vertovec (1999, 2000) and Werbner (2005) as they affect British cultural landscape values (see Tolia-Kelly, 2004, 2007a, 2007b). A focus on the value of researching the 'home' as a space of post-colonial historical narratives (see Blunt, 2003; Burton, 2003) has proven to enrich these earlier accounts, but also to politicize women's lived experiences as representing histories figured through domestic space shaped through colonial governance and landscapes. Women's attunement to 'landscapes of memory' via domestic cultures is at the heart of this research. Landscape is critical to identity formation in post-colonial terms; remembered landscapes of the past situate historical narratives about the diasporic community and are refracted in this research through material cultures in the domestic scene and on canvas.

The sites of research are threefold: the material cultures of home, the testimonies of women, and the visual images and painted canvases produced in collaboration between the artist Melanie Carvalho, the South Asian women and myself. The homes of the women in the study are considered as sites of cultural materials which operate as artefacts of a social history for British Asian diasporic cultures. The social worth of examining material cultures as archival was inspired by Appadurai (1986), Mehta and Belk (1991), Miller (1998, 2001) and Csikszentmihalyi and Rochberg-Halton (1981), with a particular focus on the often occluded cultural geographies of post-colonial women (Blunt and Rose, 1994), particularly British Asian women. British Asian Women in this book are positioned as culturally English and engaged with geographies that are set in colonial landscapes, thus challenging their usual conceptualizations as fitting cultural 'ethnic' stereotypes in academic texts (Puwar and Raghuram, 2003). The material cultures that are in the domestic space refract relationships between British Asians and various routes of migration, landscapes of home and narratives of 'heritage', community history and cultural geographies of belonging. Determining women's domestic space as a valid 'archive' is problematic (Burton, 2003) but is a necessary political challenge to accepted notions of what represents evidence for the writing of formal history. The materials of 'home' in the research offer a source of social history in the form of material cultures such as religious icons, fabrics, paintings, pictures, photographs and material objects which situate and refract biographies, and social narratives of these communities' migrant history. The textures (form, aesthetics and visual grammar) of these material cultures function as important mechanisms for signifying, mediating and presencing landscapes of belonging and identification. This is an approach to diasporic identity that centres geographical relationships with landscape as being important in mapping, figuring and identifying an English national culture that is post-colonial in its nature. These landscapes of memory are located in the spaces of colonial rule in East Africa, India, and Yemen where British and English cultural values were part of the ideological and political interventions made. They are important nodes of locating British Asian social history and represented in the research materials.

My research findings show that British Asian values of the British landscape are shaped through and interlinked with landscapes of post-colonial memory located in lived experiences under colonial rule. My collaboration with the artist Melanie Carvalho in the 'Describe a Landscape' exhibition is an attempt to materialize the complex relationship with landscape that British Asians hold. The series of paintings were produced as a result of interviews and workshops held with South Asian women in London where the women were narrating their memories of living in the colonies and the value of these landscapes to their sense of being a British citizen and culturally *English*. This series of paintings also makes tangible the women's values of landscape that are central to both their 'Britishness' and their post-colonial citizenship located beyond Britain. The expression of *Englishness* is demonstrated in visions of the lakes in Kashmir. Gardens in Mumbai are represented as 'cottage gardens'. An aspiration to living in Constable country is reflected in the narratives of the women about their expectations of living in landscapes in Britain. Cultural expressions of British citizenship in the colonies figured critically around an *English* landscape vision which did not incorporate landscapes of the colonies or those of Scotland, Ireland and Wales. This visual methodology is an original and innovative, interdisciplinary collaboration which has proven to be both a tool for multilingual research and a means to reassert different registers of landscape painting that incorporate landscape values and memories of post-colonial British citizens.

Writers on memory such as Schama (1995), Radstone (2000), Radstone and Hodgkin (2003) and Connerton (1989) offer valuable insights into social and community remembering (respectively), but these particular modes are not always available to those mobile diasporic communities, and need reconceptualizing in relation to diasporic rituals, commemoration and bodily memory practices. Diasporic citizenship is figured around cultures of 'heritage' that are multisited, transnational and in non-monumental, non-architectural mode. Therefore, the contributions of Huyssen (2003), who also offers a critical intervention on the relationships between matter and memory, and Karen Till's (2005) thesis based on memory of Berlin and its spaces of post-socialist transformation, are lacking in thinking through everyday modes of memory work for the post-colonial migrant whose relationships with monuments and materialities of public space are differently figured through exile or expulsion, non-recognition, and disenfranchisement. These sites of memory also do not include the site of 'home'. Ahmed (2000) and Fortier (2000) are critical writers in the social sciences that have disrupted the notion of 'citizenship' by figuring memory as important in considering the experience of 'racialized' minorities and the modes of multicultural citizenship and practices. However, these, for Fortier, are situated through more formalized 'collectivities' in the public sphere. Very few writers offer a post-colonial reading of 'collective memory', Avtar Brah (1999) in a short piece attempts to outline the political need to attend to race memory. In Nora (1997) and Halbwachs (1992), we are required to engage with a 'collective' but one that is a singular national collective. Legg (2005), however, does effectively critique the means through which memory is told as history and rightfully critiques these writers for their focus on particular 'collectivities' and thus particular embodied sensibilities. Sometimes what is occluded are women's memories and those figured in the relationship with the 'state' that post-colonial peoples hold. By situating the value of memory as a critical tenet of internationally mobile and racialized communities' construction of heritage, history and cultural identity, I am seeking to disrupt cultural geographies' focus on the spatiality of particularly elite cultures of landscape heritage in textual representation (Cosgrove and Daniels, 1988; Williams, 1973) and the temporality of 'bounded' citizenship that traditionally operates within Occidental or elite modes of 'tradition', 'history' and 'heritage' (see Gilroy, 1991; Hall, 1990; Samuel, 1994).

This argument is written in conversation with the material cultures literature spearheaded by the University College London Department of Anthropology, and others including Arjun Appadurai (1986), Chris Pinney (1997), Victor Buchli (2002) and Elizabeth Edwards (2001). Memory as a mobile form of locating identification temporally and spatially is valued here as a critical means of mobile and racialized communities narrating their 'heritage' and 'histories'. Through the experience of migration, formally recorded memories in the form of the museum and archive are not always amenable or inclusive to mobile diasporas. Thinking about diasporic memory on the scale of individual and collective is embedded in cultural landscapes of memory that are figured through visual and material cultures. These visual and material cultures are considered within the discourses of race and citizenship. Memories of landscape situate this social archive as connecting with international modes of citizenship and with the enforced permutations of cultural practice through intercontinental migration.

This research intends to enrich the work on the cultures of European landscape, imperial visions of landscape, landscape representation and ideologies of nationhood, and work on landscape and power – see Barbara Bender (2001), Denis Cosgrove (1984), W.J.T. Mitchell (1994) and David Matless (1998). The monograph also builds on innovative work on environmental values and visual cultures of landscape – see Burgess and Gold (1985) and Burgess et al. (1988a, 1988b) – that are integral to notions of enfranchisement, citizenship and cultures of 'being' within the environment. The monograph attends to the gaps in the literatures – between sociological and geographical writing on raceidentity – see Keith and Pile (1993), Back (1996), Dwyer (1999) and Blunt (2002) – and diasporic transnationalism – Brah (1996), Mirzeoff (2000), Vertovec (1999) and Jackson et al. (2002) – through the concept of 'landscapes of post-colonial memory'. Remembered landscapes centre identities in this book; they are negotiated and expressed through painting, material cultures and the oral narratives of the South Asian community. In this book landscape is figured as valuable beyond the usually encountered visual registers of landscape iconography; a new set of iconographies is produced on canvas in Carvalho's work. The research brings to the fore the need to reframe debates on landscape, through the lens of mobility, through an inclusive approach towards a post-colonial politics of landscape, representation and power. Here, 'British heritage' is presented as being integral to understanding post-colonial cultures of landscape in Britain; these are integral to an inclusive story of national history and heritage. Memory is important, as formal records of history do not always reflect mobile diasporas; for these communities a social narration of history is archived in oral and material cultures accumulated, circulated and gathered in the domestic landscapes of living and being. Memory and landscape are thus fused in this research, as two critical registers through which mobile, racialized, diasporic communities negotiate citizenship, heritage and narrate their history. The set of paintings produced in the research are a culmination of a project aiming to create a social archive of landscapes of memory which reflect landscapes that document an inclusive British heritage story sited beyond England.

Bibliography

Ahmed, S. (2007), 'The Language of Diversity', *Ethnic and Racial Studies* 30: 2, 235–56.

Ahmed, S. (2004a), 'Affective Economies', *Social Text* 22, 114–39.

Ahmed, S. (2004b), 'Collective Feelings: Or the Impressions Left by Others', *Theory, Culture and Society* 21, 25–42.

Ahmed, S. (2000), *Strange Encounters: Embodied Others in Post-Coloniality* (London: Routledge).

Alexander, C. (1996), *The Art of Being Black: The Creation of Black British Youth Identities* (Oxford: Clarendon Press).

Anderson, B. (1991), *Imagined Communities: Reflections on the Origin and Spread of Nationalism* (London: Verso).

Appadurai, A. (1986), *The Social Life of Things: Commodities in Cultural Perspectives* (Cambridge: Cambridge University Press).

Araeen, R. (1992a), 'Cultural Identity: Whose Problem?' *Third Text* 18, 3–5.

Araeen, R. (1992b), 'How I Discovered My Oriental Soul in the Wilderness of the West', *Third Text* 18, 86–102.

Back, L. (1996), *New Ethnicities and Urban Culture* (London: Routledge).

Bender, B. (2001), 'Landscapes on-the-Move', *Journal of Social Archaeology* 1: 1, 75–89.

Bhachu, P. (1985), *Twice Migrants: East African Sikh Settlers in Britain* (London: Tavistock).

Blunt, A. (2003), 'Collective Memory and Productive Nostalgia: Anglo-Indian Homemaking at McCluskieganj', *Environment and Planning D: Society and Space* 21, 717–38.

Blunt, A. (2002), '"Land of Our Mothers": Home, Identity and Nationality for Anglo-Indians in British India, 1919–1947', *History Workshop Journal* 54, 49–72.

Blunt, A. and McEwan, C. (2002), *Postcolonial Geographies* (New York and London: Continuum).

Blunt, A. and Rose, G. (eds) (1994), *Writing, Women and Space: Colonial and Postcolonial Geographies* (New York: Guilford Press).

Bonnett, A. (2000), *White Identities: Historical and International Perspectives* (New York: Prentice-Hall).

Bonnett, A. (1997), 'Geography, "Race" and Whiteness: Invisible Traditions and Current Challenges', *Area* 29: 3, 193–9.

Bonnett, A. (1996), '"White Studies": The Problems and Projects for a New Research Agenda', *Theory, Culture and Society* 13: 2, 145–5.

Brah, A. (1999), 'The Scent of Memory: Strangers, Our Own and Others', *Feminist Review* 61 (Spring), 4–26.

Brah, A. (1996), *Cartographies of Diaspora: Contesting Identities* (London: Routledge).

Brubaker, R. (2005), 'The "Diaspora" Diaspora', *Journal of Ethnic and Racial Studies* 28: 1, 1–19.

Buchli, V. (ed.) (2002), *The Material Culture Reader* (Oxford: Berg).

Burgess, J. and Gold, J.R. (eds) (1985), *Geography, the Media and Popular Culture* (London: Croom Helm).

Burgess, J., Harrison, C. and Limb, M. (1988a), 'Exploring Environmental Values Through the Medium of Small Groups: 1. Theory and Practice', *Environment and Planning A* 20, 309–26.

Burgess, J., Harrison, C. and Limb, M. (1988b), 'Exploring Environmental Values Through the Medium of Small Groups: 2. Theory and Practice', *Environment and Planning A* 20, 457–76.

Burton, A. (2003), *Dwelling in the Archive: Women Writing House, Home and History in Late Colonial India* (Oxford: Oxford University Press).

Clarke, N. (2002), 'The Demon-Seed: Bioinvasion as the Unsettling of Environmental Cosmopolitanism', *Theory, Culture and Society* 19, 1–2, 101–25.

Connerton, P. (1989), *How Societies Remember* (Cambridge: University of Cambridge Press).

Conradson, D. and McKay, D. (2007), 'Translocal Subjectivities: Mobility, Connection, Emotion', *Mobilities* 2: 2, 167–74.

Cosgrove, D. (1984), *Social Formation and Symbolic Landscape* (London: Croom Helm).

Cosgrove, D. and Daniels, S. (1988), *The Iconography of Landscape* (Cambridge: Cambridge University Press).

Csikszentmihalyi, M. and Rochberg-Halton, E. (1981), *The Meaning of Things: Domestic Symbols and the Self* (London: Cambridge University Press).

Dwyer, C. (1999), 'Contradictions of Community: Questions of Identity for Young British Muslim women', *Environment and Planning A* 31, 53–68.

Dyer, R. (1997), *White* (London: Routledge).

Edwards, E. (2001), *Raw Histories: Photographs, Anthropology and Museums* (Oxford: Berg).

Fortier, A.M. (2000), *Migrant Belongings: Memory, Space, Identity* (Oxford: Berg).

Gilroy, P. (1993a), *The Black Atlantic* (Cambridge, MA: Harvard University Press).

Gilroy, P. (1993b), *Small Acts: Thoughts on the Politics of Black Cultures* (London: Serpent's Tail).

Gilroy P, (1991), 'It Ain't Where You're from, It's Where You're at …: The Dialectics of Diasporic Identification', *Third Text* 13 (Winter), 3–16.

Gilroy, P. (1987), *'There Ain't No Black in the Union Jack': The Cultural Politics of Race and Nation* (London: Routledge).

Goldberg, D.T. (1993), *Racist Culture: Philosophy and the Politics of Meaning* (London: Blackwell).

Halbwachs, M. (1992), *On Collective Memory*, ed. and trans. by L.A. Coser (Chicago: University of Chicago Press).

Hall, S. (ed.) (1997), *Representation: Cultural Representations and Signifying Practices* (London: Sage, in association with the Open University Press).

Hall, S. (1996), 'New Ethnicities', in D. Morley and K.H. Chen (eds), *Stuart Hall: Critical Dialogues in Cultural Studies* (London: Routledge).

Hall, S. (1990), 'Cultural Identity and Diaspora', in J. Rutherford (ed.), *Identity, Community, Culture, Difference* (London: Lawrence and Wishart), 222–39.

Hall, S. and Du Gay, P. (eds) (1996), *Questions of Cultural Identity* (London: Sage).

Hobsbawm, E. (2005), 'The Benefits of Diaspora', *London Review of Books* 20 (October), 19–21.

Huyssen, A. (2003), *Present Pasts: Urban Palimpsests and the Politics of Memory* (Stanford, CA: Stanford University Press).

Jackson, P., Crang, P. and Dwyer, C. (eds) (2002), *Transnational Spaces* (London: Routledge).

Kalra, S., Kaur, R. and Hutnyk, J. (2005), *Diaspora and Hybridity* (London: Sage).

Keith, M. and Pile, S. (eds) (1993), *Place and the Politics of Identity* (London: Routledge).

Kitaj, R.B. (1989), *First Diasporist Manifesto* (London: Thames and Hudson).

Legg, S. (2005), 'Contesting and Surviving Memory: Space, Nation, and Nostalgia in *Les lieux de mémoire*', *Environment and Planning D: Society and Space* 23, 481–504.

Massey, D. (1993), 'Power Geometry and a Progressive Sense of Place', in J. Bird, B. Curtis, T. Putnam, G. Robertson and L. Tickner (eds), *Mapping the Futures: Local Cultures, Global Change* (London: Routledge).

Matless, D. (1998), *Landscape and Englishness* (London: Reaktion).

Mehta, R. and Belk, R.W. (1991), 'Artifacts, Identity, and Transition: Favorite Possessions of Indians and Indian Immigrants to the United States', *Journal of Consumer Research* 17 (March), 398–411.

Miller, D. (2001), *Home Possessions: Material Culture Behind Closed Doors* (Oxford: Berg).

Miller, D. (ed.) (1998), *Material Cultures* (London: University College London Press).

Mirzeoff, N. (ed.) (2000), *Diaspora and Visual Culture: Representing Africans and Jews* (London: Routledge).

Mitchell, W.J.T. (1994), *Landscape and Power* (Chicago: University of Chicago Press).

Modood, T. (1992), *Not Easy Being British* (London: Runnymede Trust).

Morrison, T. (1990), 'The Site of Memory', in R. Ferguson, M. Gever, T.T. Minh-ha and C. West (eds), *Out There: Marginalisation and Contemporary Cultures* (Cambridge, MA: MIT Press).

Natural England [website], 'R662: Audit of Non-Native Species in England', <http://naturalengland.communisis.com/NaturalEnglandShop/product. aspx?ProductID=33d69f0c-e6d5-4c21-891f-f3b23722d34e> (accessed 1 June 2006).

Neal, S. (2002), 'Rural Landscape, Representations and Racism: Examining Multicultural Citizenship and Policy-Making in the English Countryside', *Ethnic and Racial Studies* 25: 3, 442–61.

Nora, P. (1997), *Realms of Memory: Rethinking the French Past.* Vol. 2. *Traditions,* trans. A. Goldhammer (New York: Columbia University Press).

Pinney, C. (1997), *Camera Indica: The Social Life of Indian Photographs* (London: Reaktion).

Puwar, N. and Raghuram, P. (eds) (2003), *South Asian Women in the Diaspora* (Oxford: Berg).

Radstone, S. and Hodgkin, K. (eds) (2003), *Regimes of Memory* (London: Routledge).

Saad, T. and Carter, P. (2005), 'The Entwined Spaces of Race, Sex and Gender', *Gender, Place and Culture* 12: 1, 49–51.

Safran, W. (1991), 'Diasporas in Modern Societies: Myths of Homeland and Return', *Diaspora* (Spring), 83–99.

Said, E.W. (1978), *Orientalism* (Harmondsworth: Penguin).

Samuel, R. (1994), *Theatres of Memory* (London: Verso).

Schama, S. (1995), *Landscape and Memory* (London: Fontana Press).

Solomos, J. (1993), *Race and Politics in Contemporary Britain* (London: Macmillan).

Solomos, J. and Back, L. (1995), *Race Politics and Social Change* (London: Macmillan).

Spivak, G.C. (1988), 'Can the Subaltern Speak?', in C. Nelson and L. Grossberg (eds), *Marxism and the Interpretation of Culture.* Urbana, IL: University of Illinois Press.

Till, K. (2005), *The New Berlin: Memory, Politics, Place* (Minneapolis, MN: University of Minnesota Press).

Tolia-Kelly, D.P. (2007a), 'Organic Cosmopolitanism: Challenging Cultures of the Non-Native at the Burnley Millennium Arboretum', *Garden History* 35: 2, 172–84.

Tolia-Kelly, D.P. (2007b), 'Participatory Art: Capturing Spatial Vocabularies in a Collaborative Visual Methodology with Melanie Carvalho and South Asian Women in London, UK', in S. Kindon, R. Pain and M. Kesby (eds), *Participatory Action Research Approaches and Methods: Connecting People, Participation and Place* (New York and London: Routledge).

Tolia-Kelly, D.P. (2004), 'Processes of Identification: Precipitates of Re-memory in the South Asian Home', *Transactions of the Institute of British Geographers* 29, 314–29.

Turner, B. (2006), 'Citizenship and the Crisis of Multiculturalism', *Citizenship Studies*, 10: 5, 607–18.

Turner, B. (2002), 'Cosmopolitan Virtue, Globalization and Patriotism', *Theory, Culture and Society* 4: 19, 45–63.

Vertovec, S. (2000), *The Hindu Diaspora: Comparative Patterns* (London: Routledge).

Vertovec, S. (1999), 'Conceiving and Researching Transnationalism', *Ethnic and Racial Studies* 22: 2, 447–61.

Werbner, P. (2002), *Imagined Diasporas Among Manchester Muslims* (Oxford: James Currey).

Williams, R. (1977), *Marxism and Literature* (London: Oxford University Press).

Chapter 2

Intimate Distance:
Visualizing Post-Colonial *Englishness*

Introduction

Englishness is mobile. The cultural construction of England and Englishness is reliant on the scores of colonial officers, hundreds of thousands of colonial migrants, and the numerous notions and values of an 'England' that have traversed the globe with these folk and circulated with their geographical imaginaries, nostalgia and social values of nature and culture. As Young (2008) argues, Englishness itself is diasporic. This chapter is about the co-construction of Englishness through the post-colonial imaginary. A visualization of England for post-colonial migrants to Britain is foundational to the subsequent encounter with the land, people and cultural codes of being in Britain. The imagined centre of the colony is both collided with and ruptured through the post-colonial encounter; body and eye. Landscape and a visualization of a landscape of belonging is at the heart of feeling 'insitu'. Understanding our place in the world is processed through the geographical coordinates of past lived landscapes, current ones and the visualization of an ideal, enfranchising landscape of belonging. Multiple nodes of identification are resonant for migrant communities, including the multiple landscapes, natures and textures of nature that are encountered, that are visual, and that contribute to a post-colonial identity for British Asian women.

This approach, conceptually and methodologically, is a move beyond a framework of thinking identity through narrow categories of essential identities of birthplace, religion, ethnicity or even social groupings based on civic categories. What is attempted is a focus on this visual nature of the migrant conceptualization of ecological citizenship. What is privileged is an engagement with the visual iconographies of British Asian women that form their sense of identity, heritage, and landscapes of memory, in the same way in which Constable's painting *The Haywain* signifies the varieties of ecologies of an English identity based on sensibility, culture and heritage. The iconographies recorded in this research, including those of the women's former territories of lived experience, that of citizenship (as British, Indian, Kenyan and Pakistani residents) as well as sensory, visual and embodied geographies of connection and negotiation, exemplify the complex post-colonial geographies that are negotiated at all levels and scales at which they are played out. For Code (2006), what is important are understandings of a *habitability* of environment and society that are not skewed by a Western epistemic privileging of an invisible masculine figure of objective knowing; it is about *careful knowing*; 'ecological thinking is about imagining, crafting, articulating, and endeavoring to enact principles of ideal cohabitation' (p. 5). Cohabitation is about pursuing a vision of relationships between peoples, societies and ecologies of living that are equal, not the same, but where a hegemony of ideas cannot reflect the nature of the whole; acknowledging pluralities of ideas, philosophies and ecological praxis is about respecting those normally positioned as 'Other', 'Oriental' and 'outside' the violent legacies of colonial frames of thinking. Hall (1990) argues that the black body, culture and experience have been subjected to and made 'the subject of' the dominant regimes of representation, where the effects of cultural power and normalization lead '[us] to be constructed as different and other within the categories of knowledge of the West by those regimes. They had the power to make us see and experience *ourselves* as "Other"'. The reason for thinking 'visually' about post-colonial identity here is that one strand of effecting the positioning of blacks as 'Other' has been through the cultures of iconographic landscape representation. The framing of the viewer's lens is set to look upon the 'Other' and is never set to be owned, shaped, or controlled by him. The British Asian women in this research offer us the view from that positioning. If we know this visualization, a 'careful knowing' can reorientate the lens, but also enrich our ecological understandings of post-colonial citizenship, as lived, felt and memorialized. As Code (2006, p. 28) has stated:

> Ecological knowings are enacted and ecological principles derived within a transformative, interrogating, and renewing *imaginary* – a loosely integrated system of images, metaphors, tacit assumptions, ways of thinking – a guiding metaphorics that departs radically from the imaginary through and within which epistemologies of mastery are derived and enacted.

Ecological 'positioning' is inspired as an extension of Stuart Hall's (1990, p. 266) notion of 'positioning':

> Cultural identities are the points of identification, the unstable points of identification or suture, which are made, within the discourses of history and culture. Not an essence but a *positioning* Hence, there is always a politics of identity, a politics of position, which has no absolute guarantee in an unproblematic, transcendental 'law of origin'.

'Law of origin' here refers to a notion of a 'homeland', genetic, cultural and political. However, this notion of a root of all migrants and diasporic communities 'essentialises'. This chapter is entitled 'Intimate Distance' precisely because, for British Asians, there is an intimacy of engagement at the level of social connections rather than those formal connections with *Bharat* (India) through state territory and flags. The relationship is built on narratives, iconographies, memories and material memories of British Asian women with landscape – a geographical set of coordinates of identification.

The research methodology employed has enabled an ecological understanding of 'situatedness' and is designed, framed and determined through the participants' own voices and collaborative input. The 'distance' here is multiple. Primarily, it signifies the distances travelled by the women and their multiple routes through several foreign lands to arrive in Britain as 'non-native', as aliens, as 'others' of Empire and nation. The distance is also a signification of how, and is about the ways in which, landscape is valuable beyond the European frame, and how these empirical materials themselves have been alien to academic texts, which have occluded these worldly ecologies of citizenship which are here because Englishness was out there in the colonies.

What is also unravelled is the political nature of texts. The research uncovers visual, written and material texts that are positioned as such because they contain, refract and communicate the ecological situation of post-colonial identity and its geographical coordinates of memory, belonging and connection. The nature of memory in linear narratives becomes text. Souvenirs of past lives become heritage artefacts, and the iconographical nature of visual memory also becomes or forms the story of migration, or journeying to a negotiated citizenship through British territory, ending in England. These textures of mobility, or visual memories, of embodied citizenries are the 'lost' heritage in formal sites and texts that form history, heritage, cultures of display and the archive in Britain. There are textures that are foundational, and fixed points in the narrative also foundational to 'positioning' Englishness, citizenship and national memory itself. The examples given in this chapter illuminate the pathways of the visual, of memory, of landscape and post-colonial identity, where they converge at home, at the cinema, and in landscapes of Englishness.

Engaging with post-colonial cultural landscapes is critical, for both cultural cohesion, and a post-colonial understanding of Englishness. Here I write through the relationship between landscapes of post-colonial memory and British heritage and history. I focus on the constructions of Englishness as they are embedded in the national fabric. Englishness is considered here as being part of the domestic frame. This is where geographical memories of biographical landscapes that British Asian migrants hold are embedded. Collections of material fragments of things are mobilized in the diasporic journey to Britain. These solid fragments refract the memories of place, site and situated identity and through their presence in the UK landscape are a means of negotiating their British citizenship and *Englishness* as a cultural way of being. It is only on arrival in England that their way of being is seen to be discordant to the landscape and an English sensibility.

I locate my argument within the interdisciplinary literatures of historical geography, cultural geography, visual cultures, anthropology, and cultural studies. For the British Asian women that I have researched, terms such as 'heritage', 'history' and 'cultural landscape' become conceptually problematic. Cultural landscapes of Britain are at once a source of disenfranchisement, and a source of remembering landscapes of citizenship abroad. Englishness as a cultural identity is familiar for mobile, post-colonial communities living in Britain, and has been engaged with in the colonies through education, public policy and the shaping of landscape itself. British Asian memories link them to a social history forged through moving through these varied cultural landscapes, which are presented here as remembered geographical narratives that constitute heritage narratives and contribute to a hybrid citizenship of being 'British Asian'. Post-colonial time-space is presented here through a notion of memory that reflects a disrupted, fractured and fissured colonial citizenship based on residence in several continents at varying modes of colonial rule. Thus, this chapter argues that formalized notions of 'British heritage' require a re-theorization through a lens of mobility. A bounded citizenship connected to a cultural rooting to a landscape of heritage is also not available to these migrant communities, and thus there is a critical need to rethink the role of geographies of citizenship and belonging as figured through memories of 'landscape', and the narration of alternative geographies of 'heritage' and 'history' within the British Asian diaspora. This leads me into the following chapter where I outline the praxis of research which locates my methodological approach in the needs of the British Asian women's community and the theoretical literatures of participatory enquiry. Much geographical writing has focused on the iconography of landscape painting and its role in the formation of exclusionary national discourses of citizenship. This chapter is an examination of visual cultures and their role in the socio-political and geographical 'positioning' of South Asians in Britain. Visual cultures are figured as central to the processes of identity-making, and belonging, in operation in diasporic homes. They are examined as prisms through which idealized, real, imagined and iconographical landscapes of belonging are refracted. The metonymical, and multisensory nature of these visual cultures is investigated through ethnographic research with South Asian women in London.

Iconographies of British Asian Identity

In Mumbai, in 2000, Indian artist Parthiv Shah (2000) documented the Indian diasporic experience in Britain. Through an exhibition entitled *Figures, Facts and Feelings: A Direct Diasporic Dialogue*, Shah presented a photographic dialogue about the diasporic communities' positioning in relation to India – a montage of photographs of people's homes and their self-created montage of textual icons of identification with India. The reiteration of 'Indianness' through these materials is central to his recording of socio-anthropological domestic cultures. They reflect the continuing dialogue between Indians in Britain and their sense of belonging to a national territory of India. As an Indian artist observing these points of connection and creating new points for his Mumbai audience, Shah views the diaspora from within the 'motherland'. His photographs return the post-colonial gaze through which the diaspora in Britain is defined, recorded and signified, and allow for a shift in the usual flow of visual cultures from India to Britain. This 'returned-flow' raises questions about the way in which 'Indianness' is constructed by the diaspora through visual cultures and objects which, authenticated as Indian, are often part of more complicated commodity routes.

Shah engages with issues of diaspora identity from outside European art history. He works with a set of structures different from those of Black British artists, who, engaging with theories of identity, also question the political structures and prejudicial practices of art history itself.[1] Studies of visual culture are situated between art history and its 'distancing from the production of living culture, and anthropocentric concerns with the relationship between the visual and processes of subjective/cultural identification and production' (Pollock, 1996, p. 3). This intertextuality requires complex analyses and interpretation from a range of disciplines. Film, video, urban design, photography and advertising are interwoven in our everyday experience of the social world. They matter in their aesthetic and expressive nature, constituting the material realities of the political, economic, and social structures of society. Here I consider the positioning of visual cultures in the home as a means of *fixing* and *negotiating* residence in Britain for the South Asian diaspora. There is a politics to the collection and display of these cultures. They express a visual grammar and allow a 'highly individualized presencing' through their existence. I examine the role of these visual cultures in the home in the context of debate about material cultures, consumption and identity.[2]

Visual cultures are central to constructions of identity. Within cultural geography, research has been dominated by their role in discourses of landscape and national identity (Kinsman, 1995; Pollard, 1994). There has been a focus, for example, on the canvases of Turner and Constable, and on the more parochial understandings of the social and cultural values of landscape (Daniels, 1993; Matless, 1998). In these studies, landscape and Englishness have been theoretically linked as a means of contextualizing and deconstructing the function of visual imagery in the narratives of nation, and their role in the relationship between social 'structures of feeling' in the country and the city (see; Daniels, 1993; Lowenthal, 1991; Kinsman, 1997; Williams, 1993). Landscape discourse is therefore considered in the construction of geographies of belonging and being.[3] Within geography, there has been a particular focus on the effect and development of everyday landscapes and their

---

1   See, for example, Rasna Bhushan (curator), *Telling Tales: of Self, of Nation, of Art*, an exhibition held at Castle Museum and Art Gallery, Nottingham, in March–May 1998, where five Asian women artists look at post-colonial identity. Also see Roshini Kempadoo and Paloumi in the *Translocations* exhibition held at the Photographers' Gallery, London, in March–April 1997. See also Araeen, 1991; Gilroy, 1991.

2   Landscape iconography is the development of the theorization of visual culture of landscape representation, revealing political and national iconographies which inform cultural practice. Stephen Daniels (1993) contextualizes the production and consumption of these images, unravelling the multifarious historical narratives and meanings which are yielded by the visual representations he considers.

3   Tuan (1979) describes the relevance of landscape as being part of a universal desire for an 'ideal and humane habitat. ... Such a habitat must be able to support a livelihood and yet cater to our moral and aesthetic nature ... landscape allows and even encourages us to dream.'

textures on senses of place and belonging.[4] My concern is with the symbolic landscapes of nation and the local ecologies that shape our connectedness with local landscape. Richard Mabey (1980) describes these as a 'uniquely private network of meaning and association that attaches us to them' (p. 16).[5] This research examines the effect of post-colonial migration on imagined, material, and cultural landscapes of identification. Visual cultures in the home ensure a positioning of diasporic groups, through their metaphorical effect, their metonymical value and their accretion of meaning. Signifying remembered landscapes, visual cultures symbolize the experience of a nation, or reflect textures of a specific place. Colour, texture or icon within visual forms can refract memories of the experience of a different continent, a journey or simply a moment. These refracted fragments offer coordinates from which a diasporic group can *position* themselves in a dynamic connection with the past and future; the place of migration and of settlement; and the expressive cultures of these bounded territories – linked to a sense of national belonging, enfranchisement and the socio-political figuring that this allows within the new country of residence.

> It (*identity*) is not a fixed origin to which we can make some final and absolute Return. Of course it is not a phantasm either. It is something – not a mere trick of the imagination. It has histories – and histories have their real, material, and symbolic effects. The past continues to speak to us. But it no longer addresses us as a simple factual 'past', since our relation to it, like a child's relation to the mother, is always already 'after the break'. It is always constructed through memory, fantasy, narrative and myth. Cultural identities are the points of identification or suture, which are made within the discourses of history and culture. Not an essence but a positioning. Hence there is always a politics of identity, a politics of position, which has no absolute guarantee in an unproblematic, transcendental 'law of origin'. (Hall, 1990, p. 226)

My ethnographic research involved discussion groups with Asian women in North-West London, leading to individual interviews in their homes, and the analysis of transcripts and photographs (Burgess et al., 1988a, 1988b). (In this monograph, all the names of participants have been anonymized, to preserve confidentiality.) Research was grounded in the women's own accounts of the visual cultures in their homes deemed central to their sense of belonging. Here I focus on a group discussion of the Bollywood film *Guide* (1965, directed by Vijay Anand, in Hindi) and an analysis of a one-to-one interview with one of the women, Shilpa. Analysis

---

4   Humanistic geography has considered the phenomenology of *being* and *dwelling* in the landscape as a point of engaging with the experience of landscape as an intimate *home* (see Ley and Samuels, 1978; Meinig, 1979; Relph, 1976).

5   Although Mabey's argument is for ecological preservation, he situates this in the cultural meanings of local landscape ecologies, and their essential place in the well-being of human beings (see also MacNaughten and Urry, 1998, Chapter 4).

of *Guide* focuses on the positioning of its text in the memories and social practices of film-going South Asians in Britain. The values and meanings of the film text are considered in the context of a 'triangulation' process conducted by diasporic groups through cultures of identification. Triangulation, a navigational term, describes the way that a ship's coordinates can be established by measuring distance from the horizon and due north. A relational measurement, it finds location through use of fixed geographical nodes.

The interview with Shilpa allowed me to explore the value of visual cultures as they are dynamically positioned in her processes of identification and location. Born in India, Shilpa lived in Sudan and Kenya before settling permanently in London. From this interview, visual cultures in the home can be understood as multisensory. The visual becomes material in relation to identification through archival objects of heritage. This visual archive represents a set of coordinates through which triangulation, positioning and a sense of locatedness are formed and re-formed through time.

These two very different engagements with visual cultures are examples of the varied textures that signify connections with a sense of 'home', belonging and a place of identification with 'Indianness'. Discussion of the film *Guide* presents a way of understanding the social interaction that Bollywood offers the Asian diaspora, catalysing memories of events, peoples and places where the film was viewed in the past. The social act of watching therefore consolidates a cultural memory of the practices of being 'Indian', as well as triggering connections with other landscapes and ecologies memorialized in the process of migration. Exploring domestic visual cultures illustrates the very different subjective practices of everyday life, which operate through the complex aesthetic and material textures of daily living. The multisensory nature of visual cultures complicates their effect. Simultaneously, visual and material, and inscribed with a grammar of symbols, metaphor and aesthetics, they are part of an expressive cultural dialogue, as well as being culturally significant as objects.

Triangulation is the key to sustaining, creating and articulating the multiple nodes of location experienced through migration, settlement, and resettlement in the post-colonial period. The process of cultural identification relies much on the process of location, fixing positions and constructing narratives of appropriation and exclusion within the geographical space around us. Identification with these geographies through domestic visual cultures is a means of engaging with memories, landscapes and a racialized cultural connection with the diasporic network of South Asians.

Nicholas Mirzeoff (1998, p. 14) has called for an engagement with visual culture

in a far more interactive sense, concentrating on the determining role of visual culture in the wider culture to which it belongs. Such a history of visual culture would highlight those moments where the visual is contested, debated, and transformed as a constantly challenging place of social interaction and definition in terms of class, gender, sexual and racialised identities.

Engagement for South Asians concerns contestation and debate on the 'structure of feelings' which informs ideas of Indianness. These cultures of the visual resonate with aesthetics and textures of location, affecting colour, media and the cultural aesthetics of national identification. Temporal and spatial contexts of display influence the location of refracted places and fix the identity of the viewer, so that the multisensory nature of visual cultures is crucial in their role in negotiating identity for the South Asian diaspora. Cultures of the everyday are processes through which dislocation, non-identity and memory are negotiated and formed. Understanding of these cultures draws usefully on theorizations of hybridity, in-betweenness, and DuBois's concept of double-consciousness (see Bhabha, 1994, 1996; DuBois, 1989), each of which considers the post-colonial experience as a hybrid cultural identity and as a third space, a non-located and uncentred identity in flux. Double consciousness is a position of seeing oneself through an imposed 'otherness', of being in continual conflict with conscious lived experience. Expressive cultures therefore are a means of interrogating this doubleness. For example, in considering the African diaspora's relationship with musical cultures, Paul Gilroy (1993b) uses the term 'antiphony' to theorize a social memory of sound that resonates with a slave memory of the African diaspora, an aural texture that triggers the historical past and a means of reconnecting to an oral history of oppression and subjugation.

Textures of Identification

For South Asians in Britain, visual networks of culture are made possible through the development of new technologies of transport, communications and the global transmission of TV and film. The textures of memory and heritage ingrained in these visual cultures express negotiations of identity through heritage and geographical imaginings and help to counter dominant discourses of history. Archives of heritage and identity narratives are preserved in the process of self-making. Raphael Samuel's (1994) definition of heritage describes a creative process of collection, display and identification with a past landscape through the procurement of domestic objects as central to the sustenance of the self (see also Hewison, 1987; Wright, 1985). For diasporic groups, archiving through visual and material cultures preserves connections to various landscapes of migration.

Counter-culture and counter-history are produced through collections, photography and a vernacular based in the everyday, which counter insecurity in a national context – an alternative to modernity but ironically reliant on modern technology and networks of capital, transport and communications. Described by Appadurai (1997, p. 48) as 'global ethnoscapes', they constitute the new non-localized constructions of ethnic identity. These are new landscapes of group identity not based on bounded, territorialized space, and neither culturally homogeneous nor historically unself-conscious. 'The loosening of the holds between people, wealth, and territories fundamentally alters the basis of cultural reproduction' (p. 49). Through technology the imagination of many other different worlds is possible to a higher degree than ever. Fantasy fuels the way in which we imagine our potential lives, and for diasporic groups, transnational networks of cultures offer imagined utopian futures and a myriad of possible configurations in terms of identity and landscapes of belonging.

Self-making through cultures of identification is an important consideration for diasporic groups in Britain. Their heritage is not material in historical architecture, artefacts in national museums or other visual cultures of national history. Their (his)story is marginal and confined to prejudicial discourses of 'otherness' and 'primitivism'. History making is on the domestic scale; it therefore becomes a project of collecting the ephemeral (see Araeen, 1987; Gilroy, 1993b; Said, 1994). Cultural expressions of past histories or collecting artefacts cannot be carried out in a temporal scale of antiques and relics, but in a scale of decades, as a contemporary activity. Diasporic memory becomes the archive, and everyday relationships with objects become the materials of heritage. Within the politics of national heritage and history, these cultures are integral and not peripheral to an 'indigenous' national culture. Post-colonial cultures are dynamically positioned as a crucial part of the landscapes of the everyday in England, not marginal to them.

These fluid diasporic identities are created and configured through cultural imprints in the form of landscape iconography, through an embodied relationship with visuality. The eye and the body cannot be separated, and instead of individuals gazing at texts, diasporan subjects configure themselves through them. These texts could operate as nodes of belonging, but also as points of reforming identities in Britain, and so the visual becomes a practice of engaging with places, history and/ or identity. The point of engagement is privileged, rather than the unified meaning of the text, and the metonymical 'triggers of identification' become a useful way of exploring these practices – representing other places, memories and texts in a visual context of configuration. What follows is an example of the way in which the visual shifts into materiality. The value of seeing an object merges with the other sensory values of touch, hearing, taste, and smell, and practices of ocularity and embodiment are continually in flux.

Location, Journey and Post-Colonial Negations

The women in this study migrated here from a number of different countries in East Africa and the Indian subcontinent, including Kenya, Malawi, Tanzania, Uganda, Sudan, Myanmar, Pakistan, India and Bangladesh. Their routes to England parallel the migration routes within the British Empire, and as a result trace the movement of Indian labour during colonial rule in British India. Quite often for post-colonial migrants, especially those expelled from Uganda and Kenya, material possessions were given up to the state or given to workers on the farms and homes. Summer dresses specially made to make a good impression, official documents, and some small essentials were all that could be transported by the women. So for this particular group, there has been a loss, not just of locatedness, but of historical record. Migratory groups are continually omitted from history. The ethnographies of East African Asians or Indian Asians reveal the experience of in-betweenness, not just in their personal dislocation but also in their social and cultural record. No longer part of Indian history and particularly marginal to Ugandan and British histories, the materials of migration are limited in their expression of their past. Therefore cultural products such as films become prisms of location, reflecting iconographies of India, Kenya and Uganda, and allowing a connection with textual cornerstones of culture such as literature and language. Film songs are examples of this. But these texts also operate as a place through which the oral histories of migratory groups are refracted. South Asian popular culture, such as Bollywood films, has an importance not only as a cultural phenomenon in its own terms, but also as making sense of the diasporic experience of collective dislocatedness.[6] This dislocatedness was not an essential and universal experience for all those relocated through imperialism, but it was crucial for significant numbers of the South Asian population who moved to Britain in the last 60 years (Brah, 1996). For all of the women interviewed, the memory of the journey is fresh in their minds; it is a crucial point of distancing and of creating their new positionality and identities. They reference the journey as part of their development as individuals and as a point of commonality – it unites them in defining themselves as 'Asians'. The groups I recruited attracted a variety of women and, as their biographies show, their sense of connectedness was not necessarily based on language, religion or social class, but predominantly on their experience of migration from a place which, if not necessarily the same country, was positioned against being in England. At least 14 different countries are mentioned in interviews as places of origin, home or birth, constituted through memories, not necessarily as national spaces. Their combined expressions constitute a territory of culture which the women share. Experience of migration solidifies to some degree their creation of the vast place from which they came. Through memory and through refractions in countless forms in visual media, sound and sensual experiences, this place takes on emotional and psychological importance.

The meaning of 'home' is fractured and multiple for these women. Positioned in different ways, home is both a domestic and benign place where children are raised and socialized, and an oppressive space of disempowerment and subjugation. In social anthropology, home is a place where 'structures of feeling' are reinstated and evolve. Cooking, writing, talking and clothing all become crucial fixers of identification, dominated by the culture of the home as an idealized place of security and freedom and a shelter from racism and cultural oppression – a shelter from the rain. Most importantly, 'home' is a temporal and spatial fixer of belonging, security, and non-alienation, which, in their narratives, operates

---

6   'Bollywood' denotes the Bombay-based film industry, as opposed to other film production in the Indian subcontinent. Bollywood remains the most popular film culture within the South Asian diaspora.

on different scales of the domestic, local, national and international. It is within this space, simultaneously idealized and material, that the processes of making, belonging and connecting with landscapes of identification operate.

On Location with Bollywood

Bollywood film viewing at the cinema and on TV galvanizes a sense of Indianness, 'roots' and a collective experience of the diasporic journey. Film viewing activates social and cultural networks which operate as 'imagined communities' for viewers (Anderson, 1991). The women describe going to the cinema from childhood as a group experience – and the gelling of social groupings is itself sometimes more important than seeing the film. Film-going in East Africa and India is described as involving picnics and outings, mostly to open-air cinemas, and the smells, the sights, the feel of the air, the food and the society that were part of the event were central to the experience. These open-air cinemas operated in Mombassa, Nairobi, and Malawi in the 1950s, 1960s and 1970s, and the following excerpts are from women remembering those years:

> **Neela:** The effect is fantastic and I think we never used to have ordinary cinemas in Mombassa. We always used to go to these.
>
> **Anju:** It's because it's the open air, a young town watching a film together, you know ... it was like a little society.
>
> **Shazia:** It's like a picnic, but watching a movie sort of thing.
>
> **Shanta:** In Malawi we did the same ... same thing with driving. We used to do the same thing; we go early, sit there. Tell your friends if you're coming out or not. They decide we leave at the same time, we park nearby, children play, and we bring pyjamas with us, so before the picture starts we put them on.
>
> **Bhanu:** Decide what you're going to cook for each other.

The multisensory experience is clearly evident as an essential context to the film. The community watches together, as an extended family. The cinemas are outside the cityscape and reached literally by entering the jungle and crossing open land. The journey through the 'wilderness' and the exotic smell of animals are emphasized in their descriptions: one of the Indian women mocks the hyperreality of the recalled memories. The animals are larger and closer, and together the descriptions almost form a diorama of African wilderness. The auditorium becomes a film set upon which is projected the scene of watching another screen. This is just one of the ways in which present-day cinema-going is a means of connecting to previous films, film audiences and other practices of viewing.

Cinematic Dioramas

> **Manjula:** It's not pretty because where we used to go driving, we used to go through a kind of jungle. Where I am saying was always a weird smell, before we entered the drive-in side.

> **Anju:** You could hear the animals ... because of the National Parks ... and once a herd of elephants must have come out of the National Park and obviously you could hear, ah ... the noises.

> **Shazia:** These people from Africa, as the fashion progresses, their stories will get wilder and wilder! [laughing] ... next the lions will be walking in the cinema, you just wait.

> **Manjula:** ... a thousand cars coming out of the cinema. So there is always a craze, and everybody wants to get out quickly because once you pass a certain, you know, row of trees, you know it's very stinky, very bad smell because it's really outside ... the outskirts of Kampala.

In this sequence the two women are talking about two different cinemas, one on the outskirts of Kampala, and the other on the outskirts of Nairobi, but in the group discussion they conflate rather than differentiate between these two places, privileging sensual experiences over specific coordinates, places and events. Space and time are interwoven in a created memory where the specificity of smell, sound and sight is more important. These are the attributes of a place, constructed through sensory notes, memorialized in the imagination, and triggering a sensory memory that is iconographical. Cinemas in Mwanza are described as the same as those in Malawi; Mombassa is the same as Kampala in recollections of the textures of the experience. The value culture of the visual telescopes specific locations and edits reality. The tense is always mixed, past merges into present; the moment is transferable and mutable. It is nevertheless sure in its signification. Separate occasions may be conflated, but they function within a visual logic of a particular scene, a constructed landscape of cinema-going through the animal noises, dry savannah and a communal ritual. The heterogeneity of African society and 'Asianness' itself are absent from their descriptions. These memories are iconographical, reverent of an idealized landscape of the past.

Cinema-Going in the UK

When the women describe their first few weeks away from their birthplace, left behind usually through marriage, their first experiences of cinema-going in Britain are memorable. Cinema was a cultural magnet for the Asian community. It galvanized socially and empowered culturally, offering a sense of solidarity.

Seeing the physical presence of Asian bodies is an uplifting experience, reducing feelings of alienation and isolation. At the same time as altering the social geography of Britain, it awakened memories of the social geographies of East Africa and India. These new practices of going to the cinema connected with memories of past cinema-going practices. These memories were triggered by 'ethnic' trading in the cinema areas where East African and Indian delicacies, clothes, kitchen utensils, fruit and vegetables were sold alongside jewellery and cassettes of film songs.

> **Manjula:** Also the movies reminded us of home, like after coming here when we used to be reminded of home. Although I'd never been to India when I came here. I still felt as though it was part of me. I used to feel nice when I watched Indian films. You know, I felt as if I'd come from there, you know. Although I knew I wasn't from there, but my parents were. Somehow I had some kind of attachment to that place.

The women remember the films with poignancy. Tears are shed as films which have been seen over and over again in India or East Africa are reshown in Britain. But for the women it is not the text of the film that is primary, but the actual display of social and cultural aesthetics. Lalita describes her first viewing of an Indian movie after leaving India; she had been living in Montreal for nearly a year when this event happened in autumn 1976:

> **Lalita:** I said, 'What are you talking? Which one you want to see? ... Both! ... I don't want to come home. We'll have our lunch there, we'll have our dinner there and if they repeat the show I'll see that one also ... even if it's the third time'. He said, 'OK.' ... And that's what happened, they showed Kabhi Kabhie [1976, directed by Yash Chopra] first 1–4, and then 4–5 was off, and then 5–8 they showed *Bobby* and then everybody wanted Kabhi Kabhie, and then some more crowd came, and they ran a late night show 9–12 for Kabhi Kabhie and I sat again! ... Because I knew maybe we'll never see it again. It was already five months, six months, I was already married, and I hadn't seen an Indian wedding or any Indian kind of TV or anything, anything Hindi-speaking. So after that again nothing for two months and then in August they showed one movie.

The emotional response to an Indian event is significant. Lalita doesn't mind about the film shown, but enjoys the whole experience of its language and aesthetics and of being in the cinema with other Asians. Each film is three hours long and she watches three showings in a row, which demonstrates, I think, the power of the meaning of film beyond the textual. The experience is symbolic of the broader cultural practices of Indianness. Senses are stimulated in relation to a cultural appetite, encouraged through the absence of Indian aesthetics and language in the everyday. Satiation is mediated through the film text, but wholly addressed, only through the context of watching. Through cinema-going, the women are transported to sensory memories of previous cinema-goings – of a different crowd, passing comments with girlfriends, the smell of the Indian cinema – musty, old and tired:

> **Lalita:** All those things you never think of when you are there. The smell and the darkness and the romance of the cinema is different, because it was in that auditorium it was different ...

Bollywood as a Site of Memorialization Through the Film *Guide* (1965)

This cinematic socialization is the process by which the text of the film or the social activity of watching is deemed an essential part of socialization in the family (Brah, 1996; see also Gillespie, 1995). This is through religious teaching, rites and rituals, language education and simple social etiquette projected through the filmic text as well as through the process of watching. Reverencing also emerges from symbolic and iconographical imagery and narrative within the text. Cultural products such as films become sites of location, reflecting iconographies of India, Kenya, and Uganda. They enable connection with culture through literature and language, as, for example, in songs. But these texts also operate as a place through which the oral histories of migratory groups are refracted.

All of the women in the study referred to the film *Guide*, made in 1965, starring Dev Anand and Waheeda Rehman *Guide* was an enormously popular film (based on a novel by R.K. Narayan) with high production values, including an acclaimed musical score and choreography – all the ingredients necessary for a Bollywood blockbuster. The opening sequence shows landscapes which the hero, ex-tourist guide 'Raju', is traversing, shot most notably with a 'quasi-expressionistic, garish use of colour and of calendar art sets' (cited in Rajadhyaksha and Willeman, 1994, p. 96). Raju's journey is also one of the transformation of his own identity and of spiritual metamorphosis. His journey through India is figured as a particular engagement with a montage of the nation. By the end, Raju has changed into a spiritual guide and saviour, and India has become the moral landscape of his identification. Traversing the landscape cleanses him and provides him with a moral citizenship (see Matless, 1995, 1998). *Guide* demonstrates the way that a film's narrative is sometimes secondary to the visual content.[7] The film's aesthetics triggers memories of multiple experiences and oral histories for the viewing diaspora. Sometimes these are memories of memories and not necessarily of real and lived experiences (Morrison, 1990). They are symbolic of landscapes, ecological textures and even more of personal experiences. They are, for the South Asian diaspora, part of a national dialogue, between them – an imagined community of post-migration nationals – and the root source of the sense of national identity.

---

7   This is discussed in much cultural studies material; see, for example, Morley (1992).

This dialogue is a multitextual, multisensory one which shifts and fixes in different moments and different places.

The Bollywood film crystallizes processes of identification. It simultaneously offers fluidity and fixity, allowing a multidimensional relationship with images that are viewed through the eye but which contextualize the practices of the body. Indian popular commercial cinema has come to represent a kind of psychic investment for migrants from India all over the world. It operates on many different levels. For Sumita Chakravarty (1996, p. 31), 'the self enclosed romanticism of the gesture of recall, the metonymic substitution of the Hindi film for India, is generally a means of effecting closure, of constructing rigid mental boundaries between past and present, parent culture and adopted culture, belonging and exile, nationality and naturalisation.' For all the women in the research, this film *Guide* had a special meaning. For some, it was Dev's gorgeous good looks, transporting them to their teenage years. For others, it was memories of loss or of the places where freedom and sensuality were expressed. The film text operates as a prism for all these connections between spatial and temporal localities, connections which are stores of experiences and relationships.

Landscapes in these films become icons of moral order and citizenship as well as aesthetic expressions of a cultural history. The journey of Dev Anand is more than his journey between two coordinates; it is a pilgrimage through the icons of Indian landscape, which also awaken a sense of spirituality, morality and understandings of individual citizenship. For others, it holds more poignant memories.

> **Darshna:** That clip to me reminds me of how we've all moved through different places. ... And when you leave, like you've got 24 hours to leave Uganda, what do you pack in your suitcase? What do you bring with you, to this new place? You've got all your memories, all your family ... anything that's important to you in just one or two bags.

> **Shaku:** [This reminds me of] the innocent life, you know, without any worries or anything. ... Say like he hasn't got any pressure, him and his mum enjoying life ... free. ... Now, he's got his potli [knapsack – his worldly belongings] and that's it.

In the film, landscapes of Rajasthan are iconographical because they trigger visual memories of India, but not necessarily first-hand experiences. Such epics are superimposed on actual remembered journeys, memories of stories that are visualized and revered as representing a spiritual locatedness in a country that was not necessarily a place of birth. For Chakravarty (1996), *Guide* is an example of a film where impersonation operates. Raju's identity is in constant flux. He offers us an insight into the national psyche. National icons, once closely examined, are not simply bounded representations of actual places, but more political and economic collages of a particular class, consolidating a sense of national consciousness. The layers of film text offer a parallel to the workings of memory in the sense, as Toni Morrison (1990) argues that images are stores of memories for those who have no history. Memories themselves are not fixed and true, but are of stories and reflections, what she terms in her novel *Beloved* as 'memories of a rememory'.

The interview with Shilpa serves to illustrate the autobiographical and intimate narrative in which objects function as visual, metonymical devices, secured in everyday practices. These are a means of traversing the manifold connections with various 'homes', or places of belonging. The case study stands as a witness to Shilpa's biography, and the influence of its geographies on her engagement with visual cultures in the home.

Shilpa

On 28 October 1999, I interviewed Shilpa in her home. A woman with a rich biography, she was born in India and was currently living in Harlesden, North-West London. I asked her to talk about her route to England and used this to access landscapes of 'home', transition and experiences of settlement; this provided a map of formal identification. Shilpa lived in India until she was married, when she left for her husband's residence in Kenya. This marriage failed and she moved back to India.

Shilpa remarried and moved to her second husband's home in Sudan where they lived happily for four years. After the birth of their daughter, she was given the opportunity by her husband to visit her parents in India. He bought plane tickets for her and she left without her daughter for a trip 'back home'. Whilst in India, Shilpa had a visit from her brother-in-law, who informed her that the marriage was over and he handed her a suitcase with her belongings. She was consoled by her family who advised her to make a life in India. But Shilpa, then aged 27, soon made tracks back to Sudan to 'abduct' her daughter. She regained care of her daughter and fled to England where she rented a small room; from this, she was able to apply for Housing Association property. She has lived here for 15 years without going back to India, Sudan, or Kenya.

On a tour of her home, I wanted her to show me visual cultures that were important to her and her sense of 'home'. The first thing Shilpa showed me was a locked cupboard in her 'religious' room in which she kept things dear to her. She went through photographs of Kenya, some of her wedding and her husband but mostly of her daughter and their life in London.

Next, Shilpa drew out a large, old suitcase. It looked fragile; it was bamboo yellow with aluminium fittings. It was from another age and certainly another society, and she kept her treasured things in it. She reverently laid out the contents in the living room. There were pieces of jewellery and toys, attached to which was the story of her younger sister's visit to Harlesden from India about 11 years ago. Shilpa described in detail the markets in her hometown where her sister had purchased these items, its temples and shops, her father's clothing business, his sales routes through the local farms, his journeys to Kenya and Uganda, fruits and flowers, kitchens and bathrooms. Her sister had given her a model of an Indian gas stove, an open-fire design used in the back garden of her family home. Shilpa explained how she had used a similar stove in Kenya and Sudan. She described the heat in Sudan, the cold nights and the humidity.

Shilpa became very animated at the sight of a metal bucket in the bathroom which she had bought three years ago in Harlesden. She was so pleased with this find that she bought two: one to keep in case the other got damaged. She explained that she had used a bucket like this in her family home in India, and in her marital home in Sudan. It represented the morning ritual of getting up before dawn, filling the bucket with water for bathing, and setting it on a fire to heat for the whole family. Shilpa's family and life in India were evoked by this morning ritual. Since the day it was bought, she and her daughter had been using the bucket to wash in the mornings.

Through this piece of metal, Shilpa evoked many different lands and different relationships, and the textures of her family home and the environment of Sudan became implanted in Britain. The bucket became a prism through which the intimate, domestic spaces of bathing, cooking and cleaning were evoked. In these refracted cultural landscapes, were inscribed a set of aesthetics that connected Harlesden with India and Sudan. She had made reconnections with these places for her daughter; creating an identity, a citizenship, not in the national sense, but as a set of coordinates which allowed her to map her mother's biography and in turn her family and her cultures of identification. This is what the artist Kitaj (cited in Mirzoeff, 1998, p. 36) terms 'a polygot matrix'. This metaphor offers connection and infinite configurations, but within a community of collected logic.

To be able to particularize this experience as diasporic we must be able to theorize a sense of connections, as elastic and constituted through time and space, as in Raymond Williams' (1982) term, 'structures of feeling'. My premise is that these connections signify pathways to identification through their textures. Is it correct, however, to define the signposts – the suitcase and bucket – as visual culture? 'Visuality' in this research has become slippery and intangible. Sight is privileged, but it is informed by touch, smell, feel, and embodiment. In some senses a 'Westnocentric' definition of visual culture is structured around media of expression that privilege sight, each medium in turn being governed by schools of art history and 'laws' of aesthetics. But in a world of memories and landscapes, when does 'the mind's eye' get activated in relation to pieces of metal and plastic? Are the bucket and the suitcase visual keys to a past? Can the visual incorporate multi-dimensional textures that exist in everyday objects? In terms of environmental relationships and citizenship, the material and visual cultures in the home are pivotal in identity-making. They enable triangulations that allow for the process of reterritorializing in Britain.

These cultures offer a sense of inclusion that has aesthetic, sensual, and psycho-sociological dimensions. The landscapes refracted through these objects and texts of home are far from the women's place of residence and citizenship, indicating a sense of connectedness to places outside Britain at the same time as reflecting their relationship with or reaction to belonging within England. This examination of the group's relationship with visual media offers insights into what places are desired, safe and owned by this group, and which lead to the points of enfranchisement. Many of the women knew the names of plants, birds, insects, and the local ecologies of the Indian subcontinent, East Africa, Yemen and Sudan. These were intimate knowledges and connections now subsumed in their knowledges of England.

Conclusions

Through the example of the Hindi film *Guide* and the interview with Shilpa, I have illustrated the relationship between the social networks that operate through the practice of watching a film, and the connections between metaphor in the film and the metonymical character of visual cultures in the home. Through experiences of migration, sometimes enforced migration, diasporic groups experience a sense of non-identity and dislocation within the formal structures of national identity and the geographies of England. The experience of the post-colonial migrant has been the focus of much of the work by cultural theorists. I emphasize the everyday meanings and resonances of the visual cultures within the home, a core means by which diasporic groups in England make meaning of their geographical positioning. Most importantly, diasporic landscapes of belonging and identity are fixed and framed through negotiations of national cultures of citizenship, and imagined cultural geographies that are constituted as safe, secure and free. These cultures of identification are pivotal to unveiling the relationships that diasporic groups have with national landscapes, nature and environment. Cultures of the visual are prisms through which iconographies of enfranchisement, belonging, and memory are refracted. These relationships are continually dynamic, and are not an essentializing process for the definition of South Asian culture and identity. Metonymical visual signifiers resonate with the past, but are positioned in the present. The memory of the journey continues to figure within daily cultures, but exists in relation to a sense of non-identity, fracture, loss and exclusion; a relational configuration locked into the marginal position of the post-colonial subject in national dialogue.

Bibliography

Anand, V. (Director), *Guide* (1996), 183' (120').

Anderson, B. (1991), *Imagined Communities: Reflections on the Origin and Spread of Nationalism* (London: Verso).

Appadurai, A. (1997), *Modernity at Large: Cultural Dimensions of Globalisations* (Minneapolis, MN and London: University of Minnesota Press).

Araeen, R. (1991), 'The Other Immigrant: The Experiences and Achievements of Afro-Asian Artists in the Metropolis', *Third Text* 15 (Summer), 17–28.

Araeen, R. (1987), 'From Primitivism to Ethnic Arts', *Third Text* 1 (Autumn), 6–25.

Bhabha, H.K. (1996), 'Cultures in-Between', in S. Hall and P. Du Gay (eds), *Questions of Cultural Identity* (London: Sage).

Bhabha, H.K. (1994), *The Location of Culture* (London: Routledge).

Brah, A. (1996), *Cartographies of Diaspora: Contesting Identities* (London: Routledge).

Burgess, J., Harrison, C. and Limb, M. (1988a), 'Exploring Environmental Values Through the Medium of Small Groups: 1. Theory and Practice', *Environment and Planning A* 20, 309–26.

Burgess, J., Harrison, C. and Limb, M. (1988b), 'Exploring Environmental Values Through the Medium of Small Groups: 2. Theory and Practice', *Environment and Planning A* 20, 457–76.

Chakravarty, S.S. (1996), *National Identity in Indian Popular Cinema 1947–1987* (Oxford: Oxford University Press).

Chopra, Y. (director), (1976), *Khabhi, Khabie* [film].

Code, L. (2006), *Ecological Thinking: The Politics of Epistemic Location* (New York: Oxford University Press).

Daniels, S. (1993), *Fields of Vision: Landscape and National Identity in England and the United States* (Cambridge: Polity Press).

DuBois, W.E.B. (1989), *The Souls of Black Folk* (New York: Bantam).

Gillespie, M. (1995), *Television, Ethnicity and Cultural Change* (London and New York: Routledge).

Gilroy, P. (1993), *The Black Atlantic* (Cambridge, MA: Harvard University Press).

Gilroy P, (1991), 'It Ain't Where You're from, It's Where You're at …: The Dialectics of Diasporic Identification', *Third Text* 13 (Winter), 3–16.

Hall, S. (1990), 'Cultural Identity and Diaspora', in J. Rutherford (ed.), *Identity, Community, Culture, Difference* (London: Lawrence and Wishart), 222–39.

Hewison, R. (1987), *The Heritage Industry* (London: Methuen).

Kinsman, P. (1997), 'Landscape, Race, and National Identity in Contemporary Britain', unpublished PhD thesis, University of Nottingham.

Kinsman, P. (1995), 'Landscape, Race, and National Identity: The Photography of Ingrid Pollard', *Area* 27: 4, 300–10.

Ley, D. and Samuels, M.S. (eds) (1978), *Humanistic Geography* (London: Croom Helm).

Lowenthal, D. (1991), 'British National Identity and the English Landscape', *Rural History* 2, 205–30.

Mabey, R. (1980), *The Common Ground: A Place for Nature in Britain's Future* (London: Nature Conservancy Council).

MacNaughten, P. and Urry, J. (1998), *Contested Natures* (London: Sage).

Matless, D. (1998), *Landscape and Englishness* (London: Reaktion).

Matless, D. (1995), 'The Art of Right Living: Landscapes and Citizenship 1918–39', in S. Pile and N. Thrift (eds), *Mapping the Subject* (London: Routledge).

Meinig, D.W. (1979), *The Interpretation of Ordinary Landscapes* (New York: Oxford University Press).

Mirzoeff, N. (1998), *Visual Culture Reader* (London: Routledge).

Morley, D. (1992), *Television, Audiences, and Cultural Studies* (London and New York: Routledge).

Morrison, T. (1990), 'The Site of Memory', in R. Ferguson, M. Gever, T.T. Minh-ha and C. West (eds), *Out There: Marginalisation and Contemporary Cultures* (Cambridge, MA: MIT Press).

Pollard, I. (1994), *Pastoral Interlude Series, 1987–8*, researched by Kinsman P., *Renegotiating the Boundaries of Race and Citizenship: The Black Environment Network and Environmental Conservation Bodies* (Nottingham: University of Nottingham).

Pollock, G. (ed.) (1996), *Generations and Geographies in the Visual Arts* (London: Routledge).

Rajadhyaksha, A. and Willeman, P. (1994), *Encyclopaedia of Indian Cinema* (British Film Institute: Oxford University Press).

Relph, E. (1976), *Place and Placelessness* (London: Pion).

Said, E.W. (1994), *Culture and Imperialism* (London: Vintage).

Samuel, R. (1994), *Theatres of Memory* (London: Verso).

Shah, P. (ed.) (2000), *Figures, Facts, Feelings: A Direct Diasporic Dialogue* (Mumbai, India: Centre for Media and Alternative Communication).

Tuan, Y.F. (1979), 'Thought and Landscape', in D.W. Meinig (ed.), *The Interpretation of Ordinary Landscapes* (New York: Oxford University Press).

Williams, R. (1993 [1973]), *The Country and the City* (London: Hogarth).

Williams, R. (1982), *The Sociology of Culture* (New York: Schocken Books).

Wright, P. (1985), *On Living in an Old Country: The National Past in Contemporary Britain* (London: Verso).

Chapter 3
Ecological Thinking, Politics and Research Methodologies

Ecological Citizenship: Post-Colonial Landscapes, Race and Memory

It is important for me that research methods be explicit in any research dissemination, for both accountability and the contextualization of the production of knowledge. Some would view this as a pedestrian move, but I would posit this as a way of doing theory *ecologically*. Thus the conceptual mode of research should not be separate from the practice of fieldwork and dissemination. There is an ecology of practice which suffuses through the analytical mode of research, and, as such, this chapter has been included, essentially to show 'how to ...' and to situate my conclusions.

This research methodology is designed by me and is aimed at making tangible the relationships with landscape, race, post-colonial citizenships and identity that British Asian women living in London hold. It works through how these ecologies of citizenship are figured through memories, through material and visual cultures, and through a visual methodology which archives their 'ideal landscapes of home'. Landscape, race and memory are significant to their citizenship as residents of colonial lands and Britain's post-colony. Landscape as a 'way of seeing' is subverted here to incorporate a value of landscape from a non-European eye. The women involved in the research were keenly aware of the logic of landscape as a means through which to think about citizenship and identity. At first this may seem surprising, but their understandings of citizenship, belonging and national identity are learnt through a colonial perspective in the colonies (Araeen 1992a, 1992b). The category of race is a critical framework for legislation and practice in the colonies regarding their rights and access to residency and mobility in the territories. Through recording memory and the processes of memory in securing the women's connections with landscapes, their own biographies and citizenries, the research method makes tangible a record of their social history, which is triangulated through testimony, artefact and visual archive. Much of this group's history is not recorded as part of formal British history, East African history or Indian history. This research methodology is principally about creating a triangulated record of their (i) social history through oral testimonies; (ii) their cultural citizenship, as figured through their 'home(s)', through cultural materials that are visual and material; and (iii) through creating a visual archive of paintings which reflect their 'ideal' landscapes of belonging, citizenship and 'home'.

Power, Politics and Ecological Research

There is a political imperative underpinning the research design of this project. In essence, my aim was to develop a grounded research praxis that is sensitive to hierarchies of power (Massey, 1993) and is fully reflexive of both researchers' and collaborators' positionality (England, 1994). The research design has been developed in collaboration with the women who participated. In recent times, there has been a burgeoning of research based on participatory principles (e.g. Kindon et al., 2007); in all of this, ethical engagement has been a priority. The approach of all participatory researchers encompasses the assumption that the researcher is equal to participants and that a reflexive relationship with participants can yield a co-constituted set of research questions, vocabulary and disseminations. The design of this process is intended to alleviate the systemic problems of other techniques, including disruption of the researcher-centred process and analysis, and using other media aside from written text to enable a broader palette of communication, beyond verbal or written records, to make tangible cultural geographies situated in the imagination, memory and domestic scene. The effect of this is to empower and elevate the cultural meanings and value of those materials not usually central to academic research processes. Although the aim is to design a fully participatory practice, there is always present the problem of 'speaking for' or 'of' participants (see Tolia-Kelly, 2007). I aimed for a fully inclusive, empowering, enabling and transformative process, in modest and humble processes and contexts. As a participatory action researcher (PAR), I privilege the women's voices, their values and their orientations to speaking about their priorities. However, the PAR process is always contingent, and only the participants can truly identify the political and ethical hierarchies and divisions at play. I start with the assumption that we all theorize about our geographies of being and cultures of living. Our lived lives are a dialogic negotiation between ideas, matter, ideologies, and a continually evolving intellectual ecological framework for living and being. My role as a researcher thus is to enable a voice for those processes and peoples that have not been aired or engaged with, and that form part of the bodypolitic of nation and citizenship. Ecological thinking in this respect is about relocating research enquiry 'down to the ground' (Code, 2006, p. 5), or engaging with Hall's (1990) description of the continually evolving 'positioning' of the Other in cultures of citizenship that are rooted in geographies of memory, living and feeling. Here, there is a process of enquiry that engages with the habituated quotidian life of the post-colonial migrant, and considers the negotiation of landscape, belonging and citizenship. The core of positioning research enquiry in this ecological process of research is to acknowledge social categorizations as affecting mobility, territorial appropriation and framing imagined pasts and futures for living.

The identities of the women are secured through life narratives, material cultures at home, and their relationships with landscapes of belonging. These form the three strands of the research methodology. These trajectories are a triangulation of lived memories; memories concretized through material artefacts in the home; and landscape memories that situate their identity in iconographical territories of belonging. In this research, the sites where identities are made or are 'becoming' are at the heart of the investigation; memories, women's territories of home, and domestic artefacts become the evidence for social history and cultural identity. What occurs through this process is not only an elevation of memories, identities and voices of these normally 'Other' women, but also a disruption of thinking theory and methodological praxis, with the assumptions of positing the *self* of liberal tradition (encountered as *he*) most often encountered in scholarly investigations. As Code (2006, p. 203) argues, 'he has existed only in narrowly conceived theoretical places, abstracted and isolated from the exigencies and vagaries of human lives: and whenever he has figured in philosophical political theory, he has been presumptively male, usually white, privileged, able-bodied, articulate, and educated.' This political category of *self* is disrupted in feminist accounts of the stranger, self and Other (Ahmed, 2000; Ahmed et al., 2003; Haraway, 1996). However, these accounts are not usually explored through a transparent methodological practice. This research design is about taking a more transparent approach, thus embracing a synthesis between theoretical imperatives to disrupt this singular, assumed notion of *self*, and a methodology which focuses on the voices of those not usually encountered, and investigating their ecologies of citizenship, as figured at home, through memory and visual methodologies of making tangible landscapes of belonging and 'home'.

Participatory Approaches and 'Grounded Theory'

Avtar Brah's (1996) thesis argues that culture is never static, and, in Hall's (1990, p. 225) words, 'cultural identity … is a matter of "becoming" as well as "being". It belongs to the future as much as to the past.' The design of the methodology is thus a process of *triangulation* where we can triangulate coordinates of both *time* and *space* that are critical to thinking diasporic citizenship in the world. Triangulation has emerged as a theoretical geographical concept relevant to the positionality (temporal and spatial) of diasporan subjects. Temporally diasporic cultures have always been 'Other' to the West (Said, 1978) or 'primitive' in relation to the 'modern' (Hall, 1997); this is spatially and temporally. Any site of diasporic 'origin' is thus positioned as outside modernity, thus causing subjects of nations such as India, Pakistan, and the nations of East Africa to emerge as 'Other' to a contemporary Britain. What is discordant here is that the modernity of Britishness has emerged through an interdependency on 'other' lands, labours and the subjugation of other cultures. So to arrive as a post-colonial Briton, in Britain, is to seem discordant, primitive or at least 'Other' to the civilized, educated and morally complex 'native' or 'original' Briton (Gilroy, 1991). The territories and heritage of Britishness are international, and the spaces and times of diasporic identity are about both heritage and modernity, simultaneously.

Grounded theorizing is critical to the process of enquiry embraced, to achieve what May (1996, p. 87) refers to as 'theoretical adequacy', and it has been promoted by Cook and Crang (1995) in their articulation of ethnography. The method employed has ensured that grounded theorization is developed from the research data, aiming to reflect theoretical adequacy. Grounded theorization is an approach based on the work of Glaser and Strauss (1967) and Strauss and Corbin (1990). This approach politicizes the relationship between the researcher and the researched, and the collection and analysis of the research material. The grounded approach encourages reflexivity, which seeks to induce the patterns and ideas inscribed within the material collected. A researcher is then able to develop theory from the dynamic process of data collection and analysis, rather than the researchers' own preconceived theoretical convictions overshadowing the process. A full analysis was conducted with comprehensive transcriptions (Frankland and Bloor, 1999). Grounding theory in people's lives and experiences involves understanding them in a holistic way. This is the way that ideas and theories are expressed and lived in everyday life. The gathering of the research material should be situated in the social context of meaning and expression. I decided to use a participant-centred research design in the form of in-depth groups to ensure this social context in a collective forum. Given the sensitivity of the research topic and the marginalized voice of Asian women, I have chosen in-depth groups based on the model defined by Burgess et al. (1988a, 1988b). The use of small groups offers a flexible, yet ethical framework for ethnographic research, which frames the research material as it is understood in a social context. This research method is participant-centred and is committed to four principles of group research – sociality, reflexivity, liminality and empowerment, as outlined by Goss (1996a) and Burgess et al. (1988a, 1988b). These principles address the equalizing of the power dynamics between researcher and researched, and also aim to ensure that the group evolves its commitment to the research questions. This ensures meaningful participation, with the group taking joint responsibility for its work. To ensure that I addressed the politics of participation in recruiting and conducting the groups, I also drew on social anthropological methods, including PLA (participatory learning approaches) (Pretty et al., 1995), PAR (participatory action research), and feminist research methodologies (Oakley, 1981; Gluck and Patai, 1991a, 1991b; McDowell, 1992; Patai, 1991). Chiu and Knight (1999) have used group methodology in PAR to implement specific policy changes in the health service. Their experience affirms the benefits of focus group elements in attaining participation in a non-judgmental and sensitive approach. The flexibility of the group situation and the reflexivity within the group, used alongside a sensitivity to power dynamics among different race participants, have proven a good model of participatory research. The in-depth group's size and multiple meeting schedule would allow the whole group time to build up trust, and to evolve a 'voice' for all participants in an empowering process. Within the group dynamic, there is communal responsibility to be accountable and to challenge any risk to the respectful integrity of the group. I also decided to ensure additional space for individual contributions by recognizing the need for interviews with the participants at home, in 'their space'. The complex positioning of diasporic groups has been conceived as a process of triangulation which enables the figuring of the diasporan relationships with three geographical nodes. These are; (i) the actual route of migration; (ii) the place of residence and home in Britain; and (iii) the imagined landscapes of belonging. The three types of sessions have incorporated a recovery of oral histories of migration, an examination of material and visual culture as figured in the home, and the imagined (namely utopian) landscapes of home. This is the triadic relationship that has been theorized by Brah (1996) and Safran (1991). This triangulation has been conceptualized as a simple model of the diasporic networks that are operative in identity-making.

Memories of landscapes in the form of testimony, artefacts of heritage and iconographies of citizenship have been garnered. To think about memory in relation to identity formed through social history and heritage narratives, I have used Gilroy's (1993) concept of antiphony to consider visual cultures as refractive devices in the process of identification. This is multinodal identification and is explored through the way that diasporic positioning is figured through material and visual cultures. This includes the way that the materials of culture in the home operate as a means of understanding the socio-political relationship between location and national identity. Visual and material cultures are active in the practice of heritage and identity-making. Samuel (1994) advocates an inclusive model that fits well with the notion of positioning (Hall, 1990) of identity through everyday cultures. The life-worlds of the women would be the site of my group and individual discussions, as in May (1996), but these would be framed within a sociality of knowledge, attitudes and experiences recorded in a group dynamic. The lived landscape of the home environment would also allow for a further triangulation, in that the concept of home was examined in people's houses. The methodology has been informed by Raymond Williams' (1983, 1993) arguments on the *structures of feeling* which has inspired much research within cultural studies and social sciences including work within anthropology on consumption and material cultures – most notably the work of Daniel Miller (1995a, 1995b, 1998, 2001). Inevitably, this approach raises important questions about the politics of objects, the cultural geographies of home and diaspora living.

The search for participants involved mobilizing the gatekeepers in the South Asian community. I wanted to use a non-religious, more neutral social space where groups may already exist. It was rare to find women-only group meetings. I decided to recruit women from Asian advice centres because they facilitated activities that could provide ready-made women-only groups who were familiar with each other, and familiar with a secure space that the centre provided. The Asian Women's Resource Centre (AWRC) in Harlesden, North-West London, is highly politicized in its work with local authority policy and practices. The staff were vital in providing support and advice to women facing homelessness, domestic violence, enforced deportation and problems in accessing their rights. The AWRC's position in the community and its resources are exceptional, in that it is an organization effective at providing confidential, secure and supportive advice in an environment that is purpose-built for groups of women to receive information and to take part in discussion and activities. In 1998, the AWRC was awarded £250,000 lottery funding to purchase fully equipped and spacious premises in Harlesden. The Sangat Centre (SANGAT) in the London Borough of Harrow was approached because it, too, has a reputation in the (Harrow) community for good, impartial advice and support of clients. The Sangat Centre has less secure funding than the AWRC but has premises familiar to a user group, and provides a secure, confidential and supportive environment. Staff in both centres helped me to recruit two sets of women to attend an initial scoping session. From these, I recruited participants for the series of three group sessions, and an individual interview at home.

Harrow and Harlesden are places with different social-economic demographics. Asians in Harrow were described in a specially commissioned report as having the highest owner occupation ratio, and the lowest ratio of people in privately rented accommodation of all London boroughs. A report entitled *Profiling Ethnic Minority and Refugee Communities in Brent and Harrow* by the Paddington Consultancy Partnership (1999) was a source of a profile of minority communities in Harrow. In it, Harlesden residents are representative of a community highly dependent on social welfare, and a higher percentage of residents in private, rented accommodation. The decision to work in these two geographically different communities was based on understanding, in particular, the effect of social-economic positioning on individuals' enfranchisement to UK citizenship and territory. In this research project, I chose to run mixed-group sessions on the basis of the women defining themselves as Asian. I am fluent in Gujarati and have some Hindi. This encouraged me to maintain the mixed groups already active in the centres with some additional support provided by an interpreter.[1] My advertising posters were printed in four languages: Gujarati, Hindi, Urdu and Punjabi. There are debates about the advantages of mixed or homogeneous groups when conducting group research (Barbour and Kitzinger, 1999; Morgan, 1998). Also, as Chapter 4 will show, the term 'Asian' does not adequately engage with the heterogeneity of genealogies, territorialities or cultural practices that are reflected by the 'Asian women' involved.

---

1   The groups recruited at the AWRC and Sangat had a mixture of women with different language skills and ages. In terms of language barriers, although I speak Hindi and Gujarati, I employed an interpreter to assist with translation. I picked an interpreter who was a previous user of the centre but who had no current relationship with it. I thought that having two facilitators would ensure some support for me if the discussion got tense or difficult, and that it would also help with the smoothness of exchange through ensuring that all the women understood each other's contributions. In the first group meetings, I found that because the focus of the sessions was one relevant to Asian women users, it also drew out some issues for the translator. I found it much more difficult having the translator there because of her over-contributions and her need for space within the group. I found that because she had been set up as a translator the women allowed her space disproportionately to other voices because they saw her as being in a more important role than other group members. I decided therefore to run the second group without an interpreter.

This research is not an exercise in essentialising 'Asian' ethnicity of culture. Nor is it an exercise in re-evoking an impoverished social politics of 'diversity' (Ahmed, 2007). It is aimed specifically to show simultaneously the diversity of both biographies and collective social identifications that are self-defined by the women through their positioning in post-colonial times and spaces. There is a label that the research applies to the women that is partially indicated by the term 'Asian', but my conceptual aim is to reflect, in this conjoining, the social and political struggles that the women have faced. There is an empathy to oppression, to the epistemic violence that is negotiated in daily life through council forms, categorization at supermarkets, employers' interviews, public transport, immigration control and universities, to name a few. As Sara Ahmed (2007, p. 254) argues, the memory of diversity in this research is part of the project where diversity work is not only about 'accumulating the value of diversity, as a form of social currency, but also reattaching the words that embody the histories of struggle against social inequalities'. I believed that a mixed-group approach would encourage a deeper level of political and social engagement through dialogue. This is an attempt to put into practice an understanding of 'cosmopolitan sociology' (Beck, 2007; Beck and Sznaider, 2006) which rejects a seemingly benign methodological nationalism, present in research training, that continues to decentre the non-European 'Other', or post-colonial migrant. Here, the women are positioned as *cosmopolitan English*, reflecting their cultural biographies. The women recruited through the AWRC were mainly directly migrated women from India and Pakistan, and the Sangat group was dominated by 'twice-migrants' (Asians who had emigrated first to Africa and later to Britain) (Bhachu, 1985). The women defined themselves as 'Asian', and therefore I have continued with the mixed-group methodology. The interview material generated within group discussions reflects the heterogeneity of 'Asian' experience and the complexity of the processes of identification resulting from these varied intercontinental migration routes. The participants' profiles reflect the diversity of the South Asian community as it exists in London.

One area of contention in arguments about nation and national identity is the constant problem of the cultural interchangeability between a national unit, such as 'Englishness', and 'Britishness' at the scale of national citizenship. There are historic colonial reasons for this, prior to the unification of England, Scotland and Wales, and the cultural fallout from a south-centred national imaginary (Daniels, 1993). The elisions have been written through and critiqued by many others and do not need to be revisited or retheorized here (see Colley, 1992; Daniels, 1993; Gilroy, 1993; Hall, 1990; Matless, 1998). At national heritage spaces themselves, there is a constant elision between versions of nation. The elisions between English and British are maintained throughout our cultural realms; however, instead of repeating the critique, I address the notion of 'nation' as a realm of citizenry that should include all citizens who self-determine their history as British in essence. This post-colonial political distinction posits race as being important to the moral geographies of nation and national heritage; the body is continually the site of differentiation. Neal and Agyeman (2006), in their account of rural 'space such as the national parks, argue that whiteness and senses of nation are conflated through a sense of *feeling* in the countryside where 'ethnicity opens up the gaze to majoritized ethnic formations, particularly Englishness' (p. 9). So British Asians as citizens of the nation are doubly marginalized; from both cultural formations of *Englishness* and their political rights as *Britons* due to the racialized nature of belonging to constituencies of cultural inclusion, governance and equality within judicial practices.

Doing Memory, Testimony, and Biographical Landscapes

The research materials were gathered while working with groups of British Asian women over a period of 20 weeks. Other attempts at group work within the Asian community have failed (Burgess, 1996) because of the perceived reluctance of British Asian groups to commit themselves to a series of sessions, and the inability of researchers secure credibility and trust. The decision to limit the research to women was to enable women's voices, normally marginalized within and without the British Asian community, to be recorded in academic writing. Many feminist geographers have paved the way by conducting research with excluded members of society, using reflexive and empowering research methods (Barbour, 1999; Dwyer, 1999; England, 1994; Patai, 1991; Rose, 1997; Valentine, 1997; Wilkinson, 1999). Mohammed (2001) is especially eloquent on the complex positioning of the Asian researcher working with Asian women. Here, I do not claim any cultural authority, but I simply acknowledge that my positioning within the community has allowed me access and a sense of common ground in terms of diasporic identification. The nature of the group methods used to situate the women's geographical knowledge and values within their biographies allowed me to record intense, intimate, and emotional connections not normally engaged with. However, there are dangers that this situatedness could create 'separate' and 'essential' understandings of lives that are not linked dialectically to social systems of knowledge, power and lived experience. The in-depth group method in combination with a home interview effectively aimed to allow the women to become conscious creators of their own identification within a set of social spaces including their homes; this offered a material rather than an abstract context. These contexts ensured that the transcripts formed a record of the groups' social understanding and values, in situ, within their homes and community groups. The analysis of the transcripts was completed by a 'mapping' technique recommended by Burgess (1996) and Strauss (1987). It is important to note that the analysis process itself is a powerful way of oppressing, controlling and imposing order and meaning on participants' voices. This is a political act of shaping subjectivities. This 'act of violence' is described eloquently by John May (1996).

The method also allowed for individual and collective lives to be explored simultaneously. The result of recording biographies in this way has enabled a respect for individual autonomy and identity, without always collapsing results into the category of the 'social' or 'ethnic' community. This is an essential facet of ecological thinking; collapse or closure of individual identity can be determined as an epistemic denial of individual voice and values. The biographical testimonies in this first stage allowed me to develop understanding of complex values, practices and codes among mixed Hindu, Muslim, Gujarati, Pakistani and Bangladeshi groups of working-class and middle-class women, with added complications of caste differences being referred to. Within the process, I remained reflexive and my own biography was shared. The group dynamic allowed a dynamic response to the research agenda, and thus shaped it.

In the first session, I focused on the women's biographies. Each woman recounted her individual biographical routes from birth to England, to the whole group. These were personal narratives, 'oral histories' in a group forum. This setting was valuable in many ways. It was the session where the women's 'real' landscapes of home were described and recorded; a recovery of a marginalized history of South Asian migration which produced evidence of migratory routes, in relation to individual women's lifehistories. Lived landscapes, places of settlement and points of departure were critical. Mapping them showed me and the group a holistic tracing of South Asian migration and experience. The sharing of this knowledge was essential in bonding the group, and strengthening their commitment to the research project. As witnesses of each other's mapping, a level of trust and respect developed, breaking down possible religious and language barriers. In the women's centre room, I had four world maps on which I asked each woman to mark out her story, as she told it. The actual marking out exercise gave the group an understanding of each other's geographical referencing; this highlighted the scales of connections that they had with each other. It also socialized the group in each other's reference points and experiences of migration, and settlement. I used the mapping as a point to open a discussion on the shapes and textures of their lived landscapes. Individuals asked each other about the nature of society in different locations, including Sudan, Yemen, Lahore, and Bangladesh. These were on the different geographical scales of street, village, town, city, region, nation and continent. These discursive interchanges enriched the individual testimonies. Women who had lived in Kenya exchanged views with women who had family there, or had been on holiday there. Overall, a collage of the group's experiences and perceptions was recorded during the recounting of singular migration stories. I have been careful not to conflate these individual and group contributions within the analysis, but have allowed them to inform me on the women's different ways of connecting to landscapes of home. This research is reflected in Chapter 4.

Doing Material and Visual Cultures at Home

Domestic material and visual cultures, through their installation, are critical in the formation of new political identities, carving out new landscapes of belonging. These new contexts for material artefacts refigure the narration of the past imbued within them. Post-colonial memory is an important political tool, grounding both individual memory and collective cultural heritage stories. These processes are not exclusive to the South Asian population; in fact, other writers have looked at different migrant communities and their valuing of domestic artefacts as stores of cultural narratives and memorialized biographical narratives (for example, Boym, 1998; Lambert, 2001). The presence of these materials of heritage disturbs and shifts notions of Britishness. By looking at the collage of material cultures in the British Asian home as layered with aspects of memory, I have examined them as historical inscriptions within the domestic landscape. Material cultures are critical in relation to the new sites of identity-territory relations; memory-history, as I have posited it, is activated *in relation* to the new context of living. These domestic inscriptions record the post-colonial positioning that informs a politics of South Asianness within a multicultural landscape. Imbued within this political orientation is a geography of being, belonging and making home, linked directly with a post-colonial history. Within this analysis, 'multiple provenances' emerge (Parkin, 1999, p. 309), where the notion of 'home' and 'origin' are not fixed in one locus. This form of memory-history counters the unbounded notions of 'Asian' ethnicity (biological) and nationality (cultural) through a system of collective logic that is a collectively remembered and valued memory-history (Tolia-Kelly, 2004).

A second session took place at the centres, where I asked the women to bring in one or two items from their homes that were important visually to their sense of belonging. I asked them to choose the most valued items that made their place of residence their home. I described the types of objects they could bring by telling them of two things that were important to me in my home. I extended the request by saying that visual culture included photographs, videos, embroidered cloth, paintings, drawings, pictures and film posters. My aim was to encourage the women to think about everything in the home in terms of its visual importance. By bringing them into the group discussion, I wanted the women to describe the meanings of these things to each other. It was important that each person chose her own objects, evaluated their importance and described the logic of their meaning, by relating the story of each object's meaning to its aesthetics, iconic value, or role in making a home. These descriptions would be a way of elaborating previous contributions, as well as a means through which the participants could figure their sense of identity and location. In this session the women had the option of being producers. The objects that they brought were redefined by them, as being relevant to the discussion of visual culture and identity. This was intended as a session to examine the imaginative processes at play in the figuring of visual forms in the everyday world, as well as projections of the women's way of valuing different forms and images. The objects that the women brought to this session varied in type. They included wedding photos, religious icons, religious images, landscape photographs, African curios, wooden engravings, table lights, and domestic utensils. This drew me into an analysis of visual culture and the definition of visual culture itself through *memory-history* and sensory aspects of race-memory.

The multisensory nature of the engagement with these visually important objects provided a cohesive logic to both the women's cultural geographies of citizenship and the dialogic nature of cultural materials through their agentic properties.

Doing Participatory Visual Methodologies with Melanie Carvalho: 'Describe a Landscape …'

The third session was a set of art sessions. These were run in collaboration with the landscape artist Melanie Carvalho. As I thought through the focus of each group session, it became clear that I wanted the women to be involved in a creative process. To recruit an artist, I made contact with artists who were based at community centres in London, to see whether they would be interested in undertaking a collaborative session. I came across an artist who was conducting a project that seemed ideal. Melanie Carvalho had been working on issues of landscape and identity for a number of years. Her MA dissertation was entitled 'Reflections of "Home" and "Belonging" Through National Consciousness and Identity'. In it, she reflects on the relationship between national identity, landscape and memory. Intellectually, she was dealing with location and belonging, but dealing with it within the politics of national identity. On meeting Melanie Carvalho in 1997, I discovered that her current project was a London Arts Board-funded project, entitled 'Landscapes of Home'. In this project, she advertised in newspapers and magazines and distributed requests for people's descriptions of their 'ideal landscapes of home'. Her contributors were mainly artists and friends in her circle, and she wanted to attract ordinary members of local communities. In my negotiation of a research-based relationship with the artist, a mutually beneficial collaboration emerged.

Our aim in this session was to encourage these women to think about their perceptions, experiences and fantasy landscapes of home. The difficult step was to move away from a formal contribution or statement which 'fitted' their understanding of what was expected of them. We broke down some of the formalities of the group interaction by setting up a slide show. Melanie spoke about her work and her interest in landscape. She talked about landscape and the way that she used it as a device to record the different layers of connection that people had with places. Melanie used her own biography to explain the way that she had a sense of discordance with various national citizenships, and her belief that a singular sense of belonging to a land was not a possibility for most people. Melanie describes herself as British, her cultural definition is Goan, but she was born in Zambia. Through this biographical routing, she described how landscape painting is a means of inscribing all of her memories, experiences and imagined connections with places onto a single canvas. Melanie shared her own imaginative geography of home, and that helped the participants feel more comfortable about the task we set them; to write about and/or draw 'the landscape that represents your idea of home' on an A4 sheet which had space for a simple drawing or sentence.

The women were told that they could draw and/or write, and that Melanie would use these descriptions as the basis for paintings, of which a print of 'their' painting would be gifted to them. The technique was designed from original work by Claire Cooper Marcus[2] in her 'Environmental Biographies Workshop(s)'. I asked the women to take part in an exercise in visualizing the environment with eyes shut, and with minimal stimuli. After a few minutes' visualization, the women were asked to draw or write a description without talking to each other. The session was designed to capture representations of each individual's ideal geographies of home. 'Home' was used as a way into each individual's ideas of local places of dwelling and lived landscapes, but also extended into understandings of secure territories which might be signified in terminology such as 'homeland' and 'motherland'. Home is a useful analytical tool of research into diasporic connections with land on these different scales. The descriptions the women gave are a physical tracing of imagined iconographies of home. These are not necessarily formal national iconographies but are significant in the way that they resonate with the whole experience of settlement outside and within the UK. The depictions were a way of tracing imagined landscapes that were intimately bound up with the women's sense of themselves, as figured in their landscapes of belonging. The aim of gathering these on canvas was also to shift the exhibition spaces by a set of representations from a group who were normally outside the cycles of consumption and production of landscape art.

This collection of paintings have appeared in several exhibitions, including in Plymouth in January 2001, at the annual conference of the Royal Geographical Society and Institute of British Geographers. In a session entitled 'Visual Culture and Geographies of Identity', Carvalho presented a paper on her work, which has now been published in an interview format (see Anderson et al., 2000). Carvalho has curated and exhibited in several international exhibitions, including at the James Brown Galleries (New York), Cubit Gallery (London), and Hidde Van Seggelen (London), and she has published a sole-authored book entitled *Expedition* (2006). These exhibitions have proved effective in disseminating the practices of an ethnographic research project and extending the dialogue beyond the thesis production and exhibition to a forum of theoreticians and researchers.

A private view of Carvalho's portrayal of their ideas was held for the women in July 2000, a recorded session that acted as a closure to the project, in which the women received feedback from Melanie about her depictions of their spaces and places of identification. The session was also reflexive in that the participants were able to interrogate the artist. The women used this session to reflect on Melanie Carvalho's work, and her interpretation of their words and pictures of home (see Chapter 6). In the process of analysing these, ecologies of citizenship emerged:

---

2   Claire Cooper Marcus was anassociate professor at the College of Environmental Design, University of California at Berkeley. She ran environmental biography workshops in the late 1970s. I have copies of a transcript of one of these sessions given to me by Carolyn Harrison, at University College London.

nature, landscape, state, ecology and morphologies of sites, locales and places. See, in particular, the sites and modalities, as summarised in recent publications on visual methodologies (Leeuwen and Jewitt, 2001; Rose, 2001). In analysis of the descriptions, I focused on expressing their 'compositional interpretation' (Rose, 2001). Carvalho's paintings were also analysed by this method. I reviewed them in relation to Carvalho's own biography. I had interviews with all the women, including Melanie Carvalho, where they talked of certain places and landscapes which also came up in their descriptions. Carvalho's palette had been influenced by these ecologies of identity, brought together as iconographies of diaspora. My research design has been interdisciplinary. It is informed by analysis within art history, anthropology, cultural geography and cultural studies, as the subsequent chapters will show.

Some Reflections on the Process

The women attending the sessions were familiar with the resource centre. They were known to each other and used to working with a facilitator in discussion. I was working with women that had seen me at the centre before, when I had worked there in 1990–1991. Most of the women seemed relaxed and keen to help with a research project. The majority of women were dependent on welfare benefits and living in public and private rented accommodation. Their situation had an influence on the way that they related to the dynamic of group sessions, as well as their contribution to these discussions. A severe lack of confidence was initially evident in the women's response to sharing their ideas and experiences. This was not due to the environment or having a researcher present. Rather, it seemed that the women were not used to being given a voice to talk about their ideas and perceptions. The introductory scoping session and biography session were dominated by recollections of prejudice and exclusion when landscape was mentioned. Social relationships were more dominant than cultural values. This has influenced my analytical strategies, as it has drawn me to consider the relationship between their emphases on racism when considering landscape values. Experiences and feelings of exclusion were expressed at every group discussion and have been associated with every recollection of migration, settlement and cultural belonging to England.

Through the research, I learnt a lot about the participants' lives, their struggles and joys, and they learnt about my own. The researcher's identity matters in the context of the way that all individuals are marked as an ethnic group situating them within structures of social, cultural, and power positioning (Ahmed, 1996; Standfield, 1994). Although I define myself as Asian and am a woman, I have veered away from making assumptions about any essential relationship that could develop between myself and the participants, although the groups were definitely successful in achieving a reflexive, open and honest exchange between participants. My own identity and positionality were a constant consideration in the research design, in that I had decided from the beginning that I would be open, reflexive, and honest about my own biography, and contribute to the discussion. However, this was not a smooth path. The identity I shared, and the identity that the group imposed on me, were made up of different factors that contributed to the effectiveness of my breaking down barriers, and inequalities between researchers and researched. It cannot be assumed that, as an Asian woman, my identity was seamlessly compatible with the groups. This was made most obvious at lunch with the groups. I am not a vegetarian, but, as I had identified myself as being Gujarati and Kenyan, the Gujarati women in the groups were sometimes uncomfortable about the morality of my non-vegetarian diet. Eating the flesh of a dead animal made me inhumane and unprincipled. Also, the fact that I had not married a Hindu or Gujarati made me 'different', and an outsider. After my revelation of my spouse's identity, they were simultaneously polite and distant; both fascinated by this anomaly and repulsed by his 'foreignness'. This openness to disclosing and embracing situatedness (both production and consumption of) is described by Hartsock (1983, 1987) and Haraway (1990, 1997). Harvey (1992) describes the difficulties of situatedness in creating 'separate' and 'essential' lives that are not linked dialectically to social systems of knowledge, power and lived experience. This research intends to approach these situated experiences as expressive of a dialectical relationship at various scales of global geography, culture and identity politics. I have been concerned, however, with conducting this research ethically, and not to objectify, patronize or subsume the women's contributions. Feminists have written about attempts to conduct ethical research where women researching women try to eradicate the power dynamics between researcher and researched (Duelli-Klein, 1983). These are criticized by Oakley (1981) as being contradictory, as the researcher's relationship with the researched is always one of inequality. Stacey (1990) continues this critique of feminist methodologies which are assumed to be non-exploitative and collaborative. Feminist attempts at ethnographic projects which privilege the experiential, by contextualizing knowledge and inscribing the interpersonal dialogue that exists within the research process, are criticized by Stacey as deepening the level of exploitation rather than situating it in respect and reflexivity. The researcher can never invest in the research relationship with an equal measure of 'ethnographic innocence' (Stacey, 1990, p. 117), and to claim to do so implies unequal risk and dishonesty. McDowell (1992) reviews developments which attempt to assert a more egalitarian and less exploitative research design. McDowell concludes that feminist geographers need to continue to be reflexive and to construct partial, situated knowledges made from a critical position which is politicized as being one that exists in an academic power dynamic. Such a position can marginalize the research and the researcher, and thereby compromise sensitive research strategies. Rose (1997) also critiques this position in her argument, suggesting the limits inherent in attempts to be reflexive. The inequality of positioning can be further reinforced by material inequalities which dominate research dynamics in research with 'Third World women' (Patai, 1991). Her argument is that the personal, emotional revelations that occur in ethnographic research deflect from 'non-personal, institutional and political contours of the problem of material inequality' (Rose, 1997, p. 145). I believe that my project as a whole was politically ethical in that it did not seek to define the women within a bounded sense of 'Asianness' or to exoticize them as examples of the 'Other' within the metropolis.

Bibliography

Ahmed, B. (1996), 'Reflexivity, Cultural Membership and Power in the Research Situation: Tensions, Contradictions When Considering the Researcher's Role', *The British Psychological Society Psychology of Women Section Newsletter* 7, 35–40.

Ahmed, S. (2007), 'The Language of Diversity', *Ethnic and Racial Studies* 30: 2, 235–56.

Ahmed, S. (2000), *Strange Encounters: Embodied Others in Post-Coloniality* (London: Routledge).

Ahmed, S., Castaneda, C., Fortier, A.M. and Sheller, M. (eds) (2003), *Uprootings and Regroundings: Questions of Home and Migration* (Oxford: Berg).

Anderson, P., Carvalho, M. and Tolia-Kelly, D. (2000), 'Intimate Distance: Fantasy Islands and English Lakes', *Ecumene* 8: 1, 112–19.

Araeen, R. (1992a), 'Cultural Identity: Whose Problem?' *Third Text* 18, 3–5.

Araeen, R. (1992b), 'How I Discovered My Oriental Soul in the Wilderness of the West', *Third Text* 18, 86–102.

Barbour, R. (1999), 'Are Focus Groups an Appropriate Tool for Studying Organizational Change?', in R.S. Barbour and J. Kitzinger (eds), *Developing Focus Group Research: Politics, Theory and Practice* (London: Sage).

Barbour, R.S. and Kitzinger, J. (eds) (1999), *Developing Focus Group Research: Politics, Theory and Practice* (London: Sage).

Beck, U. (2007), 'The Cosmopolitan Condition: Why Methodological Nationalism Fails', *Theory, Culture and Society* 24, 286–90.

Beck, U. and Sznaider, N. (2006), 'Unpacking Cosmopolitanism for the Social Sciences: A Research Agenda', *British Journal of Sociology* Special Issue 57: 1, 1–23.

Bhachu, P. (1985), *Twice Migrants: East African Sikh Settlers in Britain* (London: Tavistock).

Boym, S. (1998), 'On Diasporic Intimacy: Ilya Kabakov's Installations and Immigrant Homes', *Critical Inquiry* 24: 2, 498–524.

Brah, A. (1996), *Cartographies of Diaspora: Contesting Identities* (London: Routledge).

Burgess, J. (1996), 'Focusing on Fear: The Use of Focus Groups in a Project for the Community Forest Unit, Countryside Commission', *Area* 28: 2, 130–35.

Burgess, J., Harrison, C. and Limb, M. (1988a), 'Exploring Environmental Values Through the Medium of Small Groups: 1. Theory and Practice', *Environment and Planning A* 20, 309–26.

Burgess, J., Harrison, C. and Limb, M. (1988b), 'Exploring Environmental Values Through the Medium of Small Groups: 2. Theory and Practice', *Environment and Planning A* 20, 457–76.

Chiu, L.F. and Knight, D. (1999), 'How Useful Are Focus Groups for Obtaining the Views of Minority Groups', in R.S. Barbour and J. Kitzinger (eds), *Developing Focus Group Research: Politics, Theory and Practice* (London: Sage).

Code, L. (2006), *Ecological Thinking: The Politics of Epistemic Location* (New York: Oxford University Press).

Colley, L. (1992), *Britons: Forging a Nation 1707–1837* (Cambridge: Yale University Press).

Cook, I. and Crang, M. (1995), *Doing Ethnographies* (Norwich: Environmental Publications).

Duelli-Klein, R. (1983), 'How to Do What We Want to Do: Thoughts About Feminist Methodology', in Gloria Bowles and Renate Duelli Klein (eds), *Theories of Women's Studies* (London: Routledge and Kegan Paul), 87–104.

Dwyer, C. (1999), 'Contradictions of Community: Questions of Identity for Young British Muslim women', *Environment and Planning A* 31, 53–68.

England, K.V.L. (1994), 'Getting Personal: Reflexivity, Positionality, and Feminist Research', *Professional Geographer* 46: 1, 80–89.

Franklin, J. and Bloor, M. (1999), 'Some Issues Arising in the Systematic Analysis of Focus Group Materials', in R.S. Barbour and J. Kitzinger (eds), *Developing Focus Group Research: Politics, Theory and Practice* (London: Sage).

Gilroy, P. (1993), *Small Acts: Thoughts on the Politics of Black Cultures* (London: Serpent's Tail).

Gilroy P, (1991), 'It Ain't Where You're from, It's Where You're at …: The Dialectics of Diasporic Identification', *Third Text* 13 (Winter), 3–16.

Glaser, B.G. and Strauss, A.L. (1967), *The Discovery of Grounded Theory: Strategies for Qualitative Research* (New York: Aldine de Gruyter).

Gluck, S. and Patai, D. (1991a), 'US Academics and Third World Women: Is Ethical Research Possible?', in S. Gluck and D. Patai (eds), *Women's Words: The Feminist Practice of Oral History* (London: Routledge).

Gluck, S.B. and Patai, D. (eds) (1991b), *Women's Words: The Feminist Practice of Oral History* (London: Routledge).

Goss, J.D. (1996a), 'Focus Groups as Alternative Research Practice: Experience with Transmigrants in Indonesia', *Area* 28: 2, 115–23.

Goss, J.D. (1996b), 'Introduction to focus groups', *Area* 28: 2, 113–14.

Hall, S. (ed.) (1997), *Representation: Cultural Representations and Signifying Practices* (London: Sage, in association with the Open University Press).

Hall, S. (1990), 'Cultural Identity and Diaspora', in J. Rutherford (ed.), *Identity, Community, Culture, Difference* (London: Lawrence and Wishart), 222–39.

Haraway, D.J. (1997), *Modest_Witness@Second_Millennium.FemaleMan©_ Meets_ Oncomouse^{TM}: Feminism and Technoscience* (New York: Routledge).

Haraway, D.J. (1996), 'Modest Witness: Feminist Defractions in Science Studies', in P. Galison and D.J. Stump (eds), *The Disunity of Science: Boundaries, Contexts, and Power* (Stanford, CA: Stanford University Press) [to be found in Chapter 3].

Haraway, D.J. (1990), *Simians, Cyborgs and Women: The Reinvention of Nature* (New York: Routledge).

Hartsock, N.C.M. (1987), 'Rethinking Modernism: Minority vs. Majority Theories', *Cultural Critique* 7, 187–206.

Hartsock, N.C.M. (1983), *Money, Sex and Power: Toward a Feminist Historical Materialism* (New York: Longman).

Harvey, D. (1992), 'Postmodern Morality Plays', *Antipode* 24: 4, 300–26.

Kindon, S., Pain, R. and Kesby, M. (eds) (2007), *Participatory Action Research Approaches and Methods: Connecting People, Participation and Place* (New York and London: Routledge).

Lambert, S. (2001), *Irish Women in Lancashire 1922–1960* (Lancaster: Lancashire Centre for North-West Studies).

Leeuwen, T.V. and Jewitt, C. (eds) (2001), *Handbook of Visual Analysis* (London: Sage).

Limb, M. and Dwyer, C. (eds) (2001), *Qualitative Methodologies for Geographers* (London: Arnold).

Massey, D. (1993), 'Power Geometry and a Progressive Sense of Place', in J. Bird, B. Curtis, T. Putnam, G. Robertson and L. Tickner (eds), *Mapping the Futures: Local Cultures, Global Change* (London: Routledge).

May, J. (1996), 'Globalisation and the Politics of Place: Place and Identity in an Inner London Neighbourhood', *Transactions of the Institute of British Geographers* 21, 194–215.

McDowell, L. (1992), 'Doing Gender: Feminism, Feminists and Research Methods in Human Geography', *Transactions of the Institute of British Geographers* 17: 4, 399–416.

Miller, D. (2001), *Home Possessions: Material Culture Behind Closed Doors* (Oxford: Berg).

Miller, D. (ed.) (1998), *Material Cultures* (London: University College London Press).

Miller, D. (ed.) (1995a), *Acknowledging Consumption: A Review of New Studies* (London: Routledge).

Miller, D. (ed.) (1995b), *Worlds Apart: Modernity Through the Prism of the Local* (London: Routledge).

Mohammad, R. (2001), '"Insiders" and/or "Outsiders": Positionality, Theory and Praxis', in M. Limb and C. Dwyer (eds), *Qualitative Methodologies for Geographers: Issues and Debates* (London: Arnold).

Morgan, D.L. (1998), *Planning Focus Groups* (Thousand Oaks, CA: Sage).

Neal, S. and Agyeman, J. (eds) (2006), *The New Countryside? Ethnicity, Nation and Exclusion in Contemporary Rural Britain* (Bristol: Policy Press).

Oakley, A. (1981), 'Interviewing Women: A Contradiction in Terms', in H. Roberts (ed.), *Doing Feminist Research* (London: Routledge and Kegan Paul).

Paddington Consultancy Partnership (1999), *Profiling Ethnic Minority and Refugee Communities in Harrow*. Report produced by the Paddington Consultancy Partnership, and published in June 1999.

Parkin, D. (1999), 'Mementoes as Transitional Objects in Human Displacement', *Journal of Material Culture* 43, 303–20.

Pretty, J.N., Guijt, I., Thomson, J. and Scoones, I. (1995), *A Trainer's Guide for Participatory Learning in Action* (London: International Institute for Environment and Development).

Rose, G. (2001), *Visual Methodologies* (London: Sage).

Rose, G. (1997), 'Situating Knowledges: Positionality, Reflexivities and Other Tactics', *Progress in Human Geography* 21: 3, 305–20.

Safran, W. (1991), 'Diasporas in Modern Societies: Myths of Homeland and Return', *Diaspora* (Spring), 83–99.

Said, E.W. (1978), *Orientalism* (Harmondsworth: Penguin).

Samuel, R. (1994), *Theatres of Memory* (London: Verso).

Standfield, J.H. (1994), 'Ethnic Modeling in Qualitative Research', in N.K. Denzin and Y.S. Lincoln (eds), *Handbook of Qualitative Research* (London: Sage).

Strauss, A.L. (1987), *Qualitative Analysis for Social Scientists* (Cambridge: Cambridge University Press).

Strauss, A.L. and Corbin, J. (1990), *Basics of Qualitative Research: Grounded Theory Procedures and Techniques* (London: Sage).

Tolia-Kelly, D.P. (2007), 'Participatory Art: Capturing Spatial Vocabularies in a Collaborative Visual Methodology with Melanie Carvalho and South Asian Women in London, UK', in S. Kindon, R. Pain and M. Kesby (eds), *Participatory Action Research Approaches and Methods: Connecting People, Participation and Place* (New York and London: Routledge).

Tolia-Kelly, D.P. (2004), 'Processes of Identification: Precipitates of Re-memory in the South Asian Home', *Transactions of the Institute of British Geographers* 29, 314–29.

Valentine, G. (1997), 'Tell Me About … Using Interviews as a Research Methodology', in R. Flowerdew and D. Martin (eds), *Methods in Human Geography: A Guide for Students Doing a Research Project* (London: Longman).

Wilkinson, S. (1999), 'How Useful Are Focus Groups for Feminist Research', in R.S. Barbour and J. Kitzinger (eds), *Developing Focus Group Research: Politics, Theory and Practice* (London: Sage).

Williams, R. (1993 [1973]), *The Country and the City* (London: Hogarth).

Williams, R. (1983), *Keywords* (London: Fontana).

Chapter 4
Diaspora Landscapes:
Mapping Post-Colonial Memory/History

Introduction

This chapter presents mappings of diasporic migration routes as evidence of the heterogeneity of these routes and the value of landscape to people's sense of identity. Research areas within 'diaspora' studies are a growing field, and challenges to traditional accounts of ethnicity, race and cultures of diaspora populations are vibrant (see especially Hopkins, 2007; Knott, 2005; Lim, 2008; Mohammad, 2001, 2007; Nayak, 2006; Phillips, 2006). This account offers insight into the social, political and cultural geographies of diaspora landscapes. This is done through a process of mapping post-colonial memory/history. The diasporic mobility of South Asians is mapped through recording oral testimonies. These collected stories of migration and citizenship based on colonial landscapes abroad are oral histories, and thus operate as a form of archive of British Asian migration. Investigated here are the geographical connections reflected in the group's biographies; thus, 'landscape' acts as a central figurative tool for situating identity, history and cultural memory. The first part of this chapter situates the case study within cultural geographies of 'landscape research', critiquing the partiality of cultural landscapes within the projects of nationalism and citizenship in Britain. These studies have marginalized or even negated the role and value of *landscape* to mobile diasporic communities, on the scale of everyday living. The research presented here shows that post-colonial communities connect with a myriad of landscapes abroad, which are critical to attaining a sense of post-colonial citizenship in Britain. 'Mapping' these connections to past landscapes offers us a process of indicating the political roots/routes of post-colonial migration and the cultural 'pasts' that are embedded in diasporic cultural citizenship in the UK. Through both the cartographic and biographical mapping of British Asian landscape connections, it is also clear that memories of other landscapes are embedded in environmental practices in Britain, and are central to social constructions of 'cultural heritage' and a testimonial to the experience of living within the colony, post-colony. The routes of migration are reflected in four maps to show the heterogeneity of landscapes experienced by British Asian women.

Post-Colonial Memory/History

This section explores testimony as a form of memory/history for post-colonial communities. Memory/history is a means to respectfully engage with those that are negated or misrepresented through epistemic violence in research and textual record. Thinking about diasporic memory as a historical record is important. Valuing 'other' modes of memory, not in the historical canon, not in the archive, and not in a medium that connects with mobile histories is part of thinking ecologically about social, cultural and political citizenship. It is in this vein that Spivak (1998) has argued that testimony, like confession, is an example of 'the genre of the subaltern, giving witness of oppression to a less oppressed other'. Testimony in this research is about bearing witness to the lived mobilities of migrant women, whose lives are made ephemeral in the project of history, negated due to their positionality. Unlike the prominent case of Guatemalan, Nobel Prize winner Rigoberta Menchú (1984), the testimonies of the women embedded here are not a singular challenge to political, colonial, cultural and epistemic violence, but are present as part of a political project of shedding light, sharing and making space for 'other' voices, reporting on Britishness, Englishness and ecologies of citizenship, in a post-colonial situation. This Latin American case of testimony and its revelation of epistemic violence and negation is, however, of consequence here. As in all British colonial territories, citizenship was legislated, and moral regulatory methods were used, but, internationally, a discordant set of identities proliferated as a result. As Castro-Gómez (2002) describes, 'No manuals were written on how to be a good peasant, a good Indian, a good black person, or a good gaucho, since all of these human types were seen as barbaric. Instead, manuals were written on how to be a "good citizen"'. For British Asians who have migrated here, aiming to be 'good citizens' of Britain has been part of a moral imperative from the beginning of their citizenship in the colonies. Their moral and cultural imperatives have been played out and compounded through double and triple migration. In the instance of Menchú's testimony, her body, her life and her academic voice were twice negated – firstly, through colonial 'regimes of truth' that framed her positioning as 'indigenous', and, secondly, through her testimony's being challenged as inaccurate and thus discredited by an academic colleague, Stoll (1999). Anthropologists have argued that this case of 'Menchú versus Stoll' is an example of a continuing imperialist ideology dominating anthropology in search of the 'complicit native' (McLaren and Pinkney-Pastrana, 2000) whose testimony does not fissure the power relation between observer and observed, colonizer and colonized. This chapter engages with making space for British Asian women's voices in their own terms.

In the process of the publication of these British Asian women's testimonies, there have been several layers of political negation. Obstacles have appeared right from the early processes of acquiring institutional approval; there were supervisory calls for my proving the legitimacy of the research, right through to the review process and editorial process of academic colleagues. Many strategies have been employed to challenge this research's credibility, academic rigour and effective representation of the women. This is of course part of academic publication, but some of it is also about not expecting the subaltern voices of these women to be legitimate, or my own research to fit within academic standards. All of these strategies have been experienced by writers who belong to the category of 'Other' (Puwar and Raghuram, 2003). These are the stalwarts of colonization, questioning the legitimacy of the project, silencing tactics, challenging claims of authenticity, and undermining one's authority. Testimony is one way to attain authority, but it remains accountable, contextually. The fallibility of memory as testimony is often attested to (Conway, 1997c). However, this research is about enriching the historical process and thus address its exclusions, in order to include and to value British Asian, post-colonial social histories. Memory/history is an important process of placing and locating people and communities both geographically and socially. Within this section, I review the value of writers on memory to enable an understanding of the 'cultural memory' and collective heritage narratives of post-colonial migrants. I write through a review of memories operating in relation to research on people who are not always part of a formal record of history. According to Connerton (1989, p. 37), reflecting on Halbwachs:

> Groups provide individuals with frameworks within which their memories are localised and memories are localised by a kind of mapping. We situate what we recollect within mental spaces provided by the group. But these mental spaces, Halbwachs insisted, always receive support from and refer back to the material spaces that particular social groups occupy.

Social scientists have been critical in thinking through the relations of memory and especially the place of nostalgia in geographical identification. These identity cultures and relationships are forged through the body, space and place (see Ahmed et al., 2003; Fortier, 2000; Legg, 2005), including diaspora societies and their complex relationship with 'home' in a post-colonial context (Tolia-Kelly, 2004). Blunt (2003) in particular examines memory in this diaspora context as active in identity formation and self-determination for those sitting 'outside' British Indian hierarchies of ethnicity and culture. Nostalgia in this context is not the commonly identified, reductive desire for 'home' through painful recollection, but one that is productive spatially and thus contributes to a creative process of making home. Productive nostalgia in this context is 'orientated towards the present and future as well as towards the past'. In fact, in the experience of shaping space and material practices within a situation, nostalgia enabled a new geopolitical possibility for Anglo-Indian women who 'belonged nowhere, neither here nor there', in Blunt's (2003, p. 724) account. For Blunt, time is displaced as the dominant parameter of defining memory; in this engagement with nostalgia, spatial, sensory and political memory is enlivened.

Addressing processes of memory/history (as I have argued) allows migrant communities to situate their identity between many coordinates, including national identity, the route of migration, and citizenship in the UK. Memory/history is a political tool to ground and resist exclusive and partial British histories. The inspiration for my concept of diasporic memory/history lies in the literature of Morrison (1990, 1994), Gilroy (1993a) and Samuel (1994). All three writers argue for the privileging of expressive cultures as nodes of an alternative history through cultures of the everyday, including narratives, musical forms and materials in the domestic landscape. These are infused with aesthetics and texturally communicate past lives, histories and a collective connection to ways of living and being that are not recorded or retrievable through the textual record of history. Memory/history is produced and sustained through everyday cultures in everyday environments. Memory/history is a creative and empowering process of self-conscious history-making as it occurs in the intimate and social landscapes of the body, the home, the street and the locale. Gilroy (1993a), Morrison (1990) and Samuel (1994) acknowledge the self-conscious history-making of those normally marginalized from the process of history writing and recording, along the lines of race, ethnicity, class and culture. These theorists, between them, extend the boundaries of what counts as artefact, collection and museum.

For South Asians, memory cultures include biographical and social memory, operating as counter-narratives to formal identity discourses (Appadurai, 1997). Viewing memory as a form of history-making is not a new project. Pierre Nora's (1984, 1986, 1989, 1996) melancholy at the loss of an organic culture of memory located in peasant life has led to the most substantive account of recording the 'realms of memory' in the French public sphere. However, the plurality of Nora's account cannot escape the burden of his occidental and singular vision of a 'national' memory situated in a material territory of France (Carrier, 2000). Nora's volumes offer a detailed analysis of the construction of French memory which is in essence an interpretation of the symbols and representations that form French commemorative history. Nora is also arguing that these commemorations are a symptom of an imperative to celebrate a *Frenchness* that is in response to a sense of national decline. Nora's definition of a national memory serves to eradicate the mobile, transnational nodes of memory-work that are at play in everyday life for post-colonial migrants, and thus, in turn, it eradicates these forms of memory/histories of those that are situated within multiple geographies of cultural remembrance, identification and biography. The cultural geographies of memory exemplify the need for continued attentiveness to the tensions within 'national memory' (Azaryahu, 2003), the tensions between national memory and 'Europeanness' (see Bialasiewicz, 2003), the nature and politics of nostalgia (Blunt, 2003), the manifestations of memory in the post-colonial world (Jazeel, 2003), and the psychoanalytical notions of individual memory processes, as exemplified in research by Martin Conway and others (Butler, 1989; Conway, 1997a, 1997b, 1997c; Gruneberg and Morris, 1978). Memory/history is a mode that is a dynamic form of historical record that meets the needs of migrant communities. In our experiences, affectual responses and encounters with memory are not necessarily racialized, but a differentiation between registers of memory and the politics of this experience is a necessary dimension to thinking and theorizing memory/history. Within its form, current British memory work (Connerton, 1989) often retains 'social memory' within a universalist frame. Racialized minorities do not figure equally within the textual record (Gilroy, 1987) or the structures of the museum archive (Hall, 2000). Chambers (1993, p. 154) also suggests that even radical historians have 'perhaps inadvertently conceded the ethical and racial pretensions of a national(ist) mythology'.

Mapping Diaspora Landscapes of South Asian Migration

The presence of South Asians in Britain results from Britain's early colonial expansion and rule and the subsequent post-colonial migration of peoples across the colonies and to the heart of colony – Britain. Within colonial narrative, the Asians are *others* within the empire and British national identity (Said, 1978). However, when they live within the heart of the metropolis (London), their presence is evidence of the multi-locatedness of Britishness and its complex identities. Under a post-colonial lens, Britishness is not only a culture of governance in India and other colonies, but also represents the national culture of British citizenship. Britishness is a mobile nationalism rendering the experience of migrant British subjects beyond the conventional paradigmatic structures of national identity. The South Asian population attains a fluid citizenry from their geographical mobility through other national landscapes governed by Britain during imperial rule. British Asians have migrated, as colonial labour and post-colonial British subjects, across several continents including Asia, Africa, Europe and North America, particularly during the nineteenth and twentieth centuries. Through their migration and settlement in new territories, South Asians have been simultaneously bound into structures of various colonial nations, territories and environments. As mobile subjects within these territories, they are and have been marginalized from inclusion in these various national cultures, resulting in their many hybrid connections with the citizenship of many nations. The community possesses an evolving consciousness of being 'post-nationals' of a colony, feeling an unfixed territorial nationalism beyond legal citizenships, and the national borders of India, East Africa and Great Britain. This creates diasporic subjectivity spatially and temporally in dynamic flux. This dynamism contrasts with the classic 'fixing' of racial identity to a singular territory within racial theory, as dominant in the colonial period, through which post-colonial diasporic communities living in the West have figured in historical, political and cultural discourse.

To understand the cultural impact of migration and settlement in Britain, it is important to explore the effect of this geographical mobility on British national identity and cultures of landscape. By examining British Asian contacts with landscapes and environments in the migratory journey, we can consider the value of other landscapes to their sense of Britishness. Mapping out South Asian migration routes and thus their multiple citizenships is the aim of this first section. The women's biographies ground the research in their lived experience and geographical connections with nations and landscapes. The second section considers the effect of this geographical mobility on securing aesthetic and material connecting points to other natures and landscapes in the home in Britain. These include organic practices in the home and environmental values expressed in domestic cultures. These are recognized here as aesthetic and textural links to past environments and landscapes. The maps produced here reaffirm the necessity for cultural theorists to continue to move beyond essentialist understandings of race, which have emerged through a colonial past (Fanon, 1959, 1961, 1967). Like other diasporas, British Asians do not figure as an essential race-group; they instead constitute a political or cultural network, in different moments and in various configurations. Many writers have examined these networks, formed through global media and communication, and the possibility of global ethnoscapes (e.g. Appadurai, 1996). Here, ethnicity is considered in its local context; grounded in the landscape of the homes of post-colonial migrant women.

In this research, the maps are formed through oral testimonies that have been recorded of the experience of migration experienced by South Asian women. Embedded within these narrations are environmental connections and attachments to nation and region, and the stories situated in lived landscapes. For the women in the study, marginalized from the national landscape of Britain (Agyeman, 1990; Malik, 1992), *other* landscapes, and ecologies become sites of affirming individual and collective identities, and points of British identity (Tolia-Kelly, 2001, 2002). These testimonies of migration are used to formulate maps of migration and descriptions which contextualize South Asian positioning within their residence in England. The inscription of these other landscapes within everyday cultures enables the valued environmental relationships, and connections to lived landscapes of the past to be recovered. In recording the variety of landscapes that British Asians have encountered in their migratory route, the maps are essential records of identification examined in considering the material cultures of gardening and the organic. As cultures which provide direct contact with territory, and soil in the home, they are important in grounding a sense of British Asian*ness*. The role of memory in this process is important in grounding a culture of landscape as being a mind's eye image of past environments.

Landscape Biographies: British Asian Women in London

The research process mapped the lives of 22 women. The social context of the migration stories led the women to talk through each other's past lives, and the places, landscapes and natures they missed, loved or simply valued. Here are included two biographies which reflect the stories behind the maps produced from the biography sessions in stage one. (This material forms the data for Figures 4.1 and 4.2.) These biographies are the lived context for the women's current geographical relationships centred in Britain, connecting them across the globe.

Beena – Born in India in 1933, Arrived in UK 1990

> **Beena:** The British have scattered our people across the whole globe. Isn't that right?

Beena's migration route is simple. She came to the UK in the 1990s from India, and her identity is embedded in remembered landscapes of India and Bangladesh. Her migration route arises from the race politics of partition and communalism in the Punjab. Beena was born in 1933, in a villagenorth-west of Punjab and Himachal Pradesh), India. As a Muslim in a Hindu state, she was vulnerable during partition. Her father was a doctor of medicine and owned a lot of land. She recalls her small hometown of Charshankar very fondly, especially the lush mango groves and banana plantations. They lived in a seven-story building owned by her father. He also owned land in Madaripur, Bangladesh, and she made many journeys with him when he visited to oversee it. Beena was married at 14, but her husband died at an early age, leaving her and her children destitute. As a woman, she had not inherited land from her father, and her husband had been a professional with no capital. She worked for a while at breaking rocks and other labour. Then she set up a business selling sugar and dry goods. To aid her family, a cousin rented her land to farm until her children came of age. In 1989, Beena became a member of parliament for Faisalbad, serving under *Benazir Bhutto* (before she became prime minister of Pakistan). When things went badly for *Bhutto*, Beena fled from Pakistan, with her whole family. She came to the UK, 10 years ago, followed eventually by her children. Beena has no means of returning to India because she is at risk of political persecution. In England, she complains of a reduced social network compounded by unemployment.

Shanta – Born in Jinja (Uganda) in 1938, Arrived London 1985

Shanta describes her closeness with Uganda as a feeling of 'being at peace'. She has memories of walking in the green, hilly landscapes, surrounded by trees and eating foods such as *mogo* (cassava). The streets were quiet with few cars. She describes a fondly remembered, idyllic childhood. After marriage, she left Jinja (Uganda) to go to Nairobi (Kenya). Shanta flew for the first time, to Malawi in 1972, when she was married. Here, the more conservative Hindu community constrained women to focus on domestic life and responsibilities. Shanta describes herself as being frightened and disempowered as a new bride entering Malawi society; her values were scrutinized on a daily basis, as were the clothes that she wore, her make-up, and the way she ran her household. The family lived in Blantyre, the suburbs to the commercial capital of Limbe. They lived opposite the president's home. Shanta describes a contented life; Blantyre had only one main shopping street, but did have the luxuries of middle-class living. She remembers quite distinctly that leisure time was really enjoyable because of their regular family picnics at Lake Malawi. She describes it as a beautiful place with the 'Thika Falls'. This is her ideal place; the gramophone is on, the children are playing in the water and food is plentiful. Shanta moved to England in 1985. On returning to Malawi recently on holiday, she found that things had changed. The lake was infected with bilharzia and the beaches were littered and ruined. She found the two-hour journey to the lake quite arduous because of the intense heat and humidity.

The journeys undertaken by these two women show the diversity of the women's experiences but also the strength of the community. The strength of connection is enough to resist erasure through the experiences of forgetting, settling, and distancing in Britain (Clifford, 1997, p. 255). The migratory journey is a point of commonality, holding them together in their description of themselves as 'Asians', but it represents the geographical breadth and scope of the diaspora. Within the group, Shanta's testimony about Lake Malawi resulted in other women recalling Victoria Falls (Zimbabwe), and Lake Nakuru (Kenya). The women shared these experiences as if they were collective. The weather, the heat and the significance of family leisure trips were factors in treating each other's recollections as similar and the same. Group discussion brings to life the significance of the migration routes in the everyday politics of being and belonging. The process of talking through these remembered territories and environments created a group collage of memories and identifications with these places and moments, which formed a shared 'territory of culture' (Tolia-Kelly, 2001). Individual stories resonated with the group as a whole, serving to map the group's commonalities through the experience of migration. Together, the women traced places of emotional and psychological importance; these were tinged with nostalgia and sometimes expressed symbolic landscapes rather than a real site of experience.

Mapping Testimonies

Figures 4.1 and 4.2 show the women's migration routes to London in a period when the British Empire was contracting. They represent the relationships between colonial subjects and the spectrum of countries controlled by the British state. The routes signify the effect of economic and labour policies of the colonial administrators between the continents of Africa, Asia and Europe, as well as the mobility of Asian males in this period. The maps show a set of active associations for the women with different landscapes through which they navigate, negotiate, and 'arrive'. In Figure 4.1, a triangular pattern is shown of travel between India to East Africa to the UK – some have migrated directly to the UK from Pakistan, Bangladesh, and India, and others from East Africa to Europe. This is very different from other 'double migrants' (Bhachu, 1985). Indian migration was limited to two or three states. At that time, the women from Pakistan and Bangladesh arrived to join men already settled here (Anwar, 1979, 1985). Within Africa, the routes were limited to Uganda, Kenya, and Malawi. (Tanzania was also a common place of residence, but is not represented in this particular sample of individuals.)

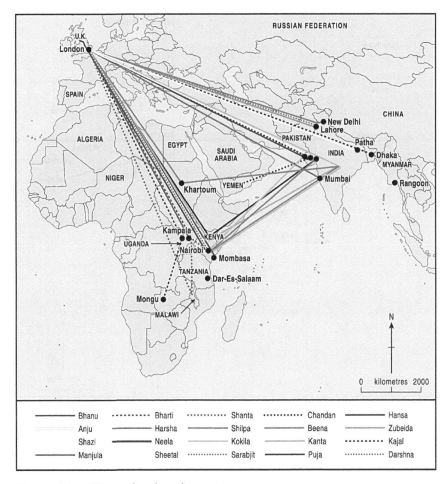

Figure 4.1   Women's migration routes

Figure 4.2 records the women's parents' migration patterns to East Africa during the 1930s and 1940s. Grandparents journeyed there from 1880 onwards, but mostly came in the 1920s and 1930s. Many parents had known a life in each of three continents and would have been part of the initial entrepreneurial business and agricultural communities established in British East Africa. The representation of fathers' migration routes shows varied settlement patterns and denotes the start of East African settlement. The women in the groups worked from memory, giving approximate dates of settlement. The role of Asian migrants at this time was to carve our new economic landscapes in East Africa. The Asians were deemed to be a source of entrepreneurship and were positioned as racial superiors to the Black African population (Brah, 1996). The politics of their race-positioning influenced the geographies of their cultural and environmental experiences and practices.

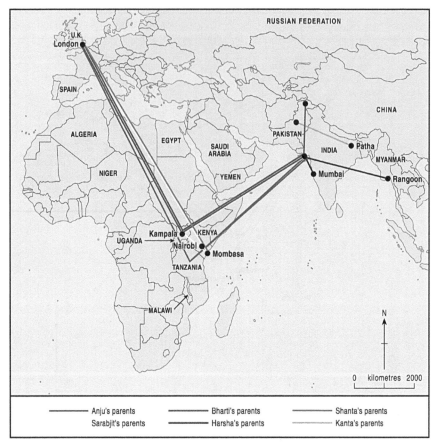

Figure 4.2   Parental migration routes

Landscapes in the colonies were shaped and figured through the particular identity-positioning of Asians within colonial discourse. These in turn influenced race-geographies, post-colony, in Britain.

'Diaspora' as a Conceptual Tool

These mappings assist in exploring the concept of diaspora itself. A singular diasporic identity and a notion of 'Asianness' itself were disrupted by these mapped routes. It is, as Hall (1990) states, that 'diaspora' is a non-essential identity, denoting a social condition and not an identity of bounded origin, ethnicity or biological singularity. The South Asian diaspora is not a singular ethnic category, and neither are the sub-sections such as *Gujarati* (Ramji, 2006). Despite the re-emergence of biologism (Gilroy, 2000), the genealogies of diasporas indicate triadic, multimodal and truly 'travelling' cultural societies (Clifford, 1997; Cohen, 1995; Safran, 1991). The writings of Paul Gilroy in particular have been crucial in my own defining of post-colonial subjectivity, and my research into the expressive cultures of South Asians, as a geographical and anthropological recording of their positioning, and into political practices of transfiguration. Gilroy (1993a) refers to understanding diasporic identity, using DuBois's (1994 [1903], p. 2) concept of double-consciousness, and this has resonance in processes of South Asian identification and appropriation as expressed in material cultures:

> After the Egyptian and the Indian, the Greek and the Roman, the Teuton and Mongolian, the Negro is the seventh son, born with a veil, and gifted with second sight in this American world – a world which yields him no true self-consciousness, but only lets him see himself through the revelation of the other world. It is a peculiar sensation, this double-consciousness, this sense of always looking at one's self through the eyes of others.

Although there are considerable differences in the shaping of Black African subjects within imperialist narratives, the cultural and racial defining of colonized peoples by skin colour continues to inform identity politics and prejudicial racist practices. The naming of 'diasporic' cultures has always been directed at the colonized and oppressed and was never attached to the colonizer. As R.J.C. Young (2008, p. 2) argues, the English themselves are a diaspora, and a cultural, ethnic diasporic identity is most recognizable, most tangible, when they are away from England itself. 'Englishness was created for the diaspora ... constructed as a translatable identity that could be adopted or appropriated anywhere by anyone who cultivated the right language, looks and culture. It then allowed a common identification with a homeland that had often never been seen.' For the colonized British, Englishness was their citizenry, and a doubleness permeated that experience. For the South Asian population living in the colonized territories, there is a tension between a lived experience of being British and the signification of identity that their non-white skin colour gives. A continued sense of doubleness is experienced. Double-consciousness has been a productive way to deconstruct the subjectivity of racial identities and to relate them to the experience of physical migration from 'home'. However, the *home* referred to here is simultaneously Asian (Indian, Pakistani, British, East-African, etc.) *and* British. The South Asian population who have migrated continue to have cultural practices that are cited as being an 'authentic cultural legacy of the "homeland"', but these are embedded with linguistic, moral, legal and geographical fluidity between being 'native' and 'non-native' both in England and during residence in the colonies. Double-consciousness allows for a deeper, progressive understanding of identity politics away from these stereotyped external signifiers, and of the lived and intimate subjectivities of South Asians living in Britain. There is a continued dialogue with the home country, but this remains a dialogue as opposed to an evolution of a syncretic bond, where one is part of the other. There is a physical, economic, and political transnationalism

(Vertovec, 1999) that informs cultural practices, but these do not together become evidence of a singular 'Indianness' or 'Asianness'. Other conceptualizations that seek to address the doubleness, or pluralities of identities that are hybrid, inbetween or bridging several territories and citizenries include the notion of translocality, especially its affective registers (Conradson and MacKay, 2007), and cosmopolitanism (Turner, 2002).

Diaspora as a conceptual tool is a productive way in which to imagine the South Asian community in Britain. It encompasses a concept of geographical positioning, and a sense of non-identity with the place of residence, but also allows for a grounded investigation into the cultural practices within the country of residence. *Diaspora* offers a sense of multinodal connections that occur simultaneously, that are directly connected to senses of identity, belonging and home (Brah, 1996; Gilroy, 1993b). Clifford (1997) argues that all diaspora discourse can only be coherent within the context of a utopian ideal. Narratives of loss and survival are driven by an understanding of the possibility of a non-oppressive, welcoming territory, either by re-territorializing and politically challenging the current state, or by maintaining an imagined homeland. Clifford (1997, p. 277) believes diaspora has shared and discrepant meanings, adjacent maps and histories. Old types of thinking or 'localizing strategies' may obscure as much as they reveal. They consider communities as bounded by an organic culture, by region or by centre and periphery. He concludes that:

> There are no post-colonial cultures or places; only moments, tactics, discourses.
> 'Post-' is always shadowed by 'neo-'. Yet 'post-colonial' does describe real, if
> incomplete, ruptures with past structures of domination, sites of struggle and
> imagined futures. Perhaps what is at stake in the historical projection of a genezia
> world or a black Atlantic is the 'prehistory of post-colonialism'.

Clifford is describing here the conflict in temporal projections of cultures. The disruption of a cultural evolution can only be considered in light of a utopian projection. This position cannot be deemed as true in light of our cultures shifting and changing in a dynamic spatial and temporal schema of a global society. In contrast to the concept of bounded national communities, diasporas are transnational. Transnationalism is an area of theoretical work focusing on the social and cultural impacts of transnational flows. Vertovec (1999), in his analysis of research in this field, outlines the different approaches to transnational movements, peoples and the value of research in this area. Drawing on Safran (1991), Vertovec (1999) defines diasporas as a social form which draws on a 'triadic relationship'. That is, where the diasporic group has a global dispersal yet a collective self-defined ethnicity, there are territorial states where such groups reside, and 'homeland' states where the group or its ancestors came from. These essentially are networks of people who form, to some extent, a self-defining public sphere with geographical imaginations about the 'homeland' or 'utopian homeland'. These networks are in the form of migration flows, or audiences for technological satellite networks, or consumers of particular goods. A group could be considered as bonded through a collective memory of heritage and national culture. The public sphere contributes to a repositioning as well as a space for the considering of location. This *placing* and positioning are not always in relation to material places; sometimes they incorporate utopian projections. Clifford (1997) in turn explores the effect of dispersal of communities on gender; he comments that 'diaspora women are caught between two patriarchies' (Clifford, 1997, p. 259). The new patriarchies are strongly reinforced as a refuge in situations where exclusion and prejudice are in operation in the host country. Community, therefore, can be a site of support, oppression and exclusion. In landscape research, the positioning of women in the discourse of landscape perspective and visual culture offers much theoretical material on their subjugation and the disempowerment of the female gaze, and, thus, on female relations with the material world.

Defining 'Asianness' Against Cultural, Racial or Ethnic Categories

The term 'South Asian' has been used in this book because it is commonly used as a geographical definition of the community living in Britain. These are the particular geographical origins of the community in Britain, which is dominated by the Indian subcontinent countries of India, Pakistan, Bangladesh, Nepal, and Sri Lanka. Due to British colonial interest in this region, these communities have subsequently had rights of citizenship and residence in the UK. Some of these early migrant groups have arrived in the UK as a secondary point of settlement after being resident in British East Africa, South Africa and a number of other, smaller settlements across the globe. The 'twice-migrants' (Bhachu, 1985) from the Indian subcontinent who have arrived via East Africa feature strongly in my research, and their definition as South Asian has raised questions of how to define the group researched. The 'twice-migrants' are African-Asians and thus do not fit neatly into the ethnic, racial or social categorization that is normally used. I have decided that the term 'South Asian' is the correct and useful term of reference for the group despite the African experience. The definition resonates with the group members themselves and has been significant in the East African group's definition of themselves whilst in East Africa. By use of the term 'South Asian' within the recruitment process, it has become clear that 'Asian' and 'Asianness' are valid when researching 'twice-migrant' communities, as their 'Africanness' is an example of the heterogeneity of the South Asian community in Britain, which is often considered as a singular ethnic community. 'South Asian' as a descriptive term for this group has been productive in terms of the geographical self-identity of the group, and is thus privileged as my definition of these migrants.

This theoretical premise of 'Asianness' has resulted in the definition of 'Asian' within the research process as being counter to ethnic, cultural or biological fixing of Asian subjectivity and lived experience. The approach taken here is part of a political act of fracturing methodologies which fix and define a static identity formed by a static culture, which, in turn, is attached to a static biological type. It is a political break away from biological and cultural essentialism. Instead, 'ethnicity', 'identity' and 'race' become processes of engagement figured through cultural materials which signify geographical mobility and idealized 'landscapes of home'. The processes of 'Asian' identity are considered as the politically dynamic operation of self-definition through social, political and cultural practices within Britain. However, this is not research in the performance of self-definition divorced from race, class and material processes, but an interrogation of how the 'positioning' of Asian identity in Britain is figured through a culture which prismatically expresses the dialectics of post-colonial race politics in Britain.

My own decision to use the term 'South Asian' in this way is consistent with contemporary commentators who state that 'Asian' as a term has 'no consistent historical or global use' (Sharma et al., 1996, p. 218). Asia is a geographically vast continent encompassing numerous nation states, a fact that renders the term 'Asian' as problematic as 'South Asian' when using it to refer to any smaller community or the origins of a national grouping. This use of the term 'Asian' in research is based on British use and context. The term 'Indian' has also been interchangeable with 'Asian' in some British contexts; however, greater public knowledge and understanding has allowed for differentiation between 'Indian', 'Pakistani', 'Bangladeshi' and 'Sri Lankan' to be made. In the US, 'Asian' refers particularly to South East Asian, Chinese, Korean or Japanese. Some cultural theorists such as Guyatri Spivak (1993, p. 54) explain that 'the name [Indian] lost specificity in the first American genocide'. Erasure has made the term unstable and irrelevant to the newer, post-colonial settlers from the Indian subcontinent. (The American context and use are relevant in that I have referenced theories of African-American writers such as Toni Morrison and bell hooks, who focus on a Black experience that is African-American.) The term 'Indian' within a British context refers predominantly to 'Hindu' migrants, as opposed to 'Pakistani', who are labelled 'Muslim'; these definitions are neither accurate nor useful in defining this South Asian group. Modood (1992; Modood and Werbner, 1997), however, argue that 'ethnicity' is the most important framework needed to define minority groups. For him, the Asian community is figured through religious practice, and definitions such as 'Muslim' are privileged as ethnic categories over geographical definitions such as 'Asian', 'Indian' and 'Pakistani'. Modood chooses to privilege religious identity because he believes that the Muslim groups face political, social and economic oppression. Modood considers these discriminations as part of a broader analysis of the hierarchy which exists within the British South Asian community, leading him to identify a 'Muslim underclass'. Modood (1992, p. 43) breaks down distinctions within the Asian community crudely as material and cultural differences between 'achievers' and 'believers'. Although belief forms the forefront of cultural oppression, and the resulting material and social reality, a definition based on religion as indicating ethnicity becomes reductive despite its attempt to recognize cultural prejudice. This treats the group as having a static, bounded ethnicity which is also fixed as such by prejudicial commentators and writers. Bobby Sayyid (1997) describes signification based on religious practice as a continued orientalist approach to defining Muslims living within Europe. Sayyid's 'Islamism' is a theorizing of Muslim fundamentalist activity, based on a political self-determination deriving from religious belief, as a counter-narrative to orientalism. Islamism then becomes a 'project which attempts to transform Islam from a nodal point in discourses of Muslim communities into a master signifier' (Sayyid, 1997, p. 48). Sayyid's research is predominantly a counter-orientalist research, but it does reveal that a solely ethnic definition is disingenuous about the group's own experiences and practices but also instrumental in polarizing the debates. As Claire Alexander (2000) also argues, such ethno-religious definitions are a remaking of bounded and absolute identities.

In the research, I sought a politicized identification that incorporated the South Asian group as both politically Black and culturally Asian, but not limited to these over-determined accounts of ethnicity which are commonplace in academic writing (Sharma et al., 1996, p. 3). According to Sharma et al. (1996), music is the expressive cultural production through which a new sociology of cultural studies can be positioned. These authors are against the exoticization that can occur through researching *difference*. Their work applies a code of ethics to their research practice, which is 'against simplification, against anti-politics, against victimologies' (Sharma et al., 1996, Introduction). The political definition of Black is most pertinent when dealing with a theorization of identity that considers a political and cultural materialist account of identity. This is the position from which I identify the dynamics of cultural and political identity that is an important component of expressive and lived identities. 'Black' as a political term is one which signified the united oppressed minorities in the anti-racist movement in Britain after the 1960s.

The terms 'Black' and 'Asian' do not have to be mutually exclusive, unless you consider American race politics where Black does mean specifically of African-Caribbean origin. This difference between American and British terminology is a symptom of the different anti-racist movements in Britain and the US. The anti-racist movement in Britain has always been led by united Asian and African-Caribbean activists in a movement that has been dominated by the slogan 'Black and White unite and fight'. In the US, however, the legacy of Martin Luther King and the 1960s civil rights movement has been dominated by the politics of Black nationalism, as described by Gilroy (1987, 1994). However, cultural critics such as Tariq Madood (1994) and Stuart Hall (1995) believe that the term 'Black' in contemporary race politics holds less resonance and is in decline. There is continued discussion in contemporary literature of the value and definition of 'Black' as a term of identity (Alexander, 1996; Gilroy, 1987, 1994; Sharma et al., 1996), and it is quite clear that the term can be as reductionist as it can be inclusionary, and it has been a tool of oppressive discourse as well as uniting an anti-racism based on culture and biological prejudice. The value of 'Black' as a political term has been to position cultural theory within race politics. In historical and contemporary analysis the disciplinary research subject of racialized 'other' which has formed the basis of socio-anthropological disciplinary analysis can no longer be reconstituted through cultural, religious or biological essentialisms but ensures that the politics of race are central to the analysis. A cultural vignette of an 'ethnic' community is no longer possible without an analysis of the racism experienced by the subjective identities within that community being redrawn without reference to the power of colonialist orientalism.

When we deal with expressive cultures, it is important to examine them in the context of the political, social and economic relationships within which these cultures are embedded. By using cultural materialism as an entrypoint into understanding the value of expressive cultures, Sharma et al. (1996, p. 7) decide that it is essential to continue with considering Black as a political identity that needs re-contextualizing and defining:

> A premature closure of this debate is liable to reproduce the historical amnesia of the post-modern condition. The valency of 'Black' is a political positionality that has strategically united disparate groups against increasingly organised and vicious manifestations of Euro-racism.

In my research, I have valued Toni Morrison precisely because she locates expressive cultures within a politicized and material geography of Black identity. Morrison's Black identity refers to an exclusive African-American experience. Her writing is described as 'irrevocably Black'. This presents a problem of the transferability of Morrison's theories on memory, race and culture to the Asian experience in the UK. This resulted in my search for an 'irrevocably Asian' writer who expounds theories of memories, race and culture. This is a not an anti-essentialist position. Morrison treats the expressive culture of literature in its power to evoke and create dialogue about identities of the present being linked to a past colonialism that has ruptured Black subjectivity through oppression and subjugation. The value of expressive cultures in resonating and opening up a dialogue which becomes dialectically positioned across spatial and temporal zones along the axis of a racialized experience is exactly what I wish to examine. Morrison theorizes Blackness not as an essential experience despite her exclusive definition of Black. She offers a dialectical basis from which to view identity, and within this memory and culture are fused as the space of unhindered cultural cohesion through spatial and temporal experience. This is a move away from a recording, measuring and defining of cultural texts and products; instead they are considered as crucial points of dynamic dialogue within and without the community. A public sphere is also materialized amongst the members of a racialized community in relation to their geographical context. Kaur and Kalra (1996, Chapter 9) re-examine this view of culture in their theorizing of *Br-Asian* and *Transl-Asian*, where Britain remains at the core of their notion of transcontinental expressive cultural network:

However this is not a fixed centre but one in oscillation with other centres ...
[it is an] arena of cultural flows, not entirely geographically grounded, not
always nationally bounded, but constantly on the move. (Kalra and Kalra, 1996,
p. 224)

Within my research, the geography of expressive cultures of British South Asians
is framed by a notion of *diaspora*. Diaspora is productive in that it offers a
framework for these spatial configurations which are connected to specific points
of reference within the Asian community in Britain. This is not an essential notion
of 'origin' and 'home' but a way of looking at routes that are the basis of the Asian
communities' common identification. The migratory experience from the same
nodal points of post-colonial South Asian migration offers a way of understanding
their positionality here in Britain. The concept of 'circulation' and 'flows' may
have been an alternative to viewing the group as a diaspora; however, the research
does make explicit heterogeneity and difference in the origins of the individuals
included within this definition of South Asian. The theorizing of identity by
the concepts of 'circulation' and 'flow' does not easily allow for a set of fixed
nodal points relating to colonial history and the related contemporary points of
cultural intercourse. Diaspora locates specifically a framework from which I have
mapped counter-origins, idealized homes and unfixed routes of cultural integrity
that are beyond biological and culturally essentialist understandings of diasporic
communities.

Social Memory and British Asian Identities

Whilst living in Britain, the women's current lives are dialectically linked to
their past environments as indicated in the maps above. Connections with these
environments are woven into the fabric of everyday life. The following section
of discussion reveals how their environmental memories of Asia and East Africa
operate in their senses of being here.

Shazia: When I go back home, I become what I used to be.

Darshna: What else is there apart from your family?

Shazia: It's the air ... It's the smell. I mean that ... For me ... that's a ... maybe
you all go there and think it's all so dirty, it's so foggy. I get asthma, I get this,
I get that ... but for me you know the heat, the smell, when I get off the plane I
feel ahhh ... When I come back, I tell Lalit [her husband], 'There's no noise on
the road, there's no people on the road ... it's just like a dead place – you can
ride for miles and not see another human being, whereas there, you step out of
the house and there are millions of people all around you.

Here Shazia is talking through a sense of remembering a connection with the physical landscape of India. In contrast, the landscape in Britain is alienating; the air is cold, and the streets less densely populated. Her comparison reflects a physical alienation from the British social landscape, and the air itself is underlined with a sense of marginality from social intercourse on the street. For these groups, the preserving of social memory becomes a political act. Remembering the landscape and the air itself contributes to a grounded sense of self, located within memories of past environments. Social memory acts as an adhesive force, sustained through rituals and everyday cultural practices. These sensory triggers in the everyday make a past territory tangible. Individual recollections often have resonance amongst the whole group. Social memory in the South Asian diaspora is constitutive of a collage of stories, embedded in environments and landscapes, thus forming a 'territory' of cultural history and identity. In the absence of formal historical inclusion, memorialization and heritage in this form are increasingly part of the imaginative realm and are maintained through the sensual, iconographic, textural, and aural signifiers within domestic visual and material cultures. Textures in the home become connecting points back to past landscapes, as well as being a source of reflection of stories about Asian history and landscapes. Here, the textures referenced include soil, plants and childhood gardens, which are often sources of links to landscapes abroad and a medium for sustaining a sense of self-identity.

Memory, Landscape and *Placing*

Psychologists have considered the location of memory in our construction of self and have traced the mechanisms and processes of memory within the context of life stages, trauma and socialization (Conway, 1997a). Memory frames and folds into our contact with environments in a myriad of ways, sustaining our sense of the past, fracturing our sense of place now, and offering a continuous source of dialogue between multiple space-times. Memory is effectively part of the landscape (Lowenthal and Prince, 1965; Lowenthal, 1979, 1985); however, it sits as a dynamic and powerful tool for the creation, sustenance and disruption of our sense of self within everyday geographies. Texturally, the scent, sound, taste, and texture of memory contextualize our body experiences within past experiences of dwelling, environment and places of being. Home for migrant groups is a transitory experience; therefore, a site-specific study of memories located in the space of homemaking, and enfranchisement in the territory of Britain, allows for the study of memories which operate as significant in the creation of a rooting and belonging in Britain for British Asians. By considering the dynamic force of memory as a *placing* mechanism, we can establish a conceptual frame for understanding post-colonial identity as it is figured through place, nature and landscape. Textures in the home signify memory and can be interrogated as a store of cultural coordinates of actual routes of identification for the British

Asian diaspora. The biographical migration flows, mapped in the first section, are critical in understanding geographical connections that relate to the negotiation of modern diasporic identities. The process of *placing* is involved in the figuring of identity for migrants. 'Placing' has been theorized by phenomenological and humanist geographers (Relph, 1976) to address the way that we *live* and *be* within an environment, including sensory engagements with the environment. Placing through the senses offers us a matrix of textures through how we situate ourselves and in turn are ourselves *positioned.* Sensory memory thickens these matrices of sensory engagement by the presence of other time-spaces that assist in our being, dwelling and identification with place, home, and landscape. Sensory memories, as expressed here, are not simply contexts to memory, but are dialogic connections with the dialectics of post-colonial social memory as expressed in contemporary environments.

This dialectical relationship between memory and *placing* is critical for diasporic groups who have migrated through varied landscapes, but who have also varied political connections with territory and national identity. By using the refractory qualities that materials of cultures offer, I have attempted to *place* post-colonial migrants within a matrix of valued connections that actively connect individuals with place, natures and landscape. The site of a red rose cultivated in a domestic garden in Britain can uncover the geographical coordinates through which identity is constituted. A rose can refract a memory of a past home or a connection to a symbolized sense of past national affiliation to Kenya, but it also can be central in a present home in securing a sense of British Asianness (see Tolia-Kelly, 2001). Manjula, below, describes her memory of the East African landscape as signified by a rose bush. The loss of connection with the rose reflects the distance and loss of contact with the gardens left behind. Remembering this black rose, links her to a collage of narratives about the planting practices, the colours and the landscape of plantations in Africa:

> **Manjula:** We had a beautiful house in Kampala. We had huge gardens all round. We had so many guava trees and mango trees and all. I still remember my sister sent me some rose bushes from Nairobi which I grew in the front garden, and one of them was a black one. Black Rose. Just when it was coming into bud, I had to come here to do my studies. I thought when I go back I'll see my garden, I was so looking forward to seeing it again, with the new lawn and the new roses.

For many migrant communities, material possessions are lost or left behind in the process of migration. The effect of this loss means that any preserved cultural articles in the UK are imbued with an enhanced meaning and value. As remnants of the past home, or bought as souvenirs after settlement in Britain, these have a significant role in the process of *location, belonging* and *placing* for this group.

Neela: It's just nice to have something from the past ... mostly, we left things as they were. Because all you could bring were suitcases. I remember the house keys and the shop keys – we gave it at the police station, and the car keys we gave it to the driver. So it was like you were passing it on. ... We gave a few things away because we knew we couldn't bring much. ... Still in my mind I believed it was temporary because I couldn't see another life for my parents starting anew ... I thought, so there would always be a home [in Kenya] but it never happened. ... You know what, it was so traumatic, I've really tried to block that out. I remember snatches but I really blocked it out of my mind. ... We didn't have anything.

The women's parents left countries where they had built their lives and businesses, from clearing land to developing new markets and building economies. For some of the families, the shock was too much to bear. Individuals dealt with this in different ways. For three of the women's fathers, Uganda became something that the household did not talk about. For others the strategy was to believe themselves to be moving to the UK only temporarily.

Sensory Memories and Landscape

There are a myriad of sensual memories we carry around with us that resonate with past environments, people and events. Scents, sounds, tastes, aesthetics and textures are evoked through remembering the past. These body-memories (Edgerton, 1995) are refracted through contact with materials in the present. They operate as a gateway into other environments, moments and social experiences. A sphere of sensual references that reflect past events and sensory textures is recalled when talking through the practices of gardening and contact with soil, plants, flowers and food. Thus, sensory memories trigger a collective connectedness to a 'territory of culture' which is shared within a collective group of South Asians who have travelled separately in their journey to Britain. The multisensory nature of these connections is an important element of the geographical cultures of being (MacNaughten and Urry, 1998; Urry, 1990) and contribute to processes of identification with place, environment and landscapes of belonging. The memories of past environments are mobile and transportable (Bohlin, 1998; Lovell, 1998) as a *portmanteau* (Crosby, 1986) of cultural memory. This is where domestic cultures create a set of historical narratives which act as both a store of refractive memorials of past stories and gateways to body-memories.

For many of the women, the effect of migration is evident in their attitudes and relationships with nature and climate; the heat of the sun, the dryness of the air and the brightness of the sky are vivid memories as migration has taken them out of this ecological context. The environmental differences between equatorial cities in the African savannah, dry tropical forests, and Indian desert and coastal biomes are evidenced in their descriptions. Arrival and adjustment are just as much about adjustment to climate as they are about making a home in a new nation. The shift in physical environment influences them psychologically and physically; the weather is a feature of their isolation and a feature of the obstacles they have to overcome to sustain day-to-day living. The women's ecological engagements highlight an orientation to particular environmental textures. A set of ecological aesthetics can be traced through their connectedness. The women's recollections include those intimate connections with kitchen-gardens and more formal national landscapes. Their narratives describe real, imaginary and symbolic relationships with natural textures. These memories are transplanted to the practices of cultivating gardens in their homes in Britain. Often, this important process of rooting replicates the practices of planting abroad in places such as Kenya, India and Uganda.

These recollections re-create and import 'other' environments into England, but also re-contextualize organic icons of life in another place. These narratives express the complicatedness of migration and dislocation, but they can also reflect the pleasure that the women get from plants, trees and landscape in England. English roses are reminders of roses in Uganda, or fuchsias trigger memories of bougainvilleas in Africa and India. In the testimonies of the women, certain plants, trees and textures recur – papaya, guava, mango, palm, bougainvillea, jasmine, gardenia, and hibiscus flowers, and the colour of the soil are particular landscape icons which are metonyms for other continental ecologies. These reflect the experience of growing up in the Indian subcontinent or Africa prior to living in London, as well their active engagement with English plants and flowers in England and abroad, for, quite often, plants abroad were introduced by the British. The women in the group cultivated African and Indian species too. In East Africa, gardening was not only a pleasurable pastime, but was also very important to the women's sense of their feminine role in contributing to the household larder, as well as being considered creative and aesthetically improving to the fabric of the home. This is a custom of ensuring a culture of fecundity in the home landscape. Below is an example of how the ecology and environment are recalled; some gardens are no longer actively cultivated, but are neglected. The women try to re-create these remembered ecological textures in their own gardens in England. When I visited the women's homes, the women talked of attempts at re-creating past gardens by growing palms, guava, sweetcorn and flowers. They were planting things that they engaged with in their past homes; this process was evidence of a need to cultivate a new landscape in Britain relevant to their experience of migration.

The texture, colour and smell of soils are also remembered by the women in their narratives as having properties not characteristic of the soil here. This is a romantic consideration, a romanticized earth. Below is an example of how places are conflated; the soil of Kenya then becomes a memory of Uganda. Shazia's memory is of Kenya's rich fertile soil, which has a special scent when wet; on her memory of this, Bharti is transported back to Uganda. There is an elevation, a reverence and a feeling of the sublime, which is repeated in the other women's experiences:

Shazia: I think when I went to Kenya about three years ago. I can't remember, and the red soil. It reminded me straightaway of the soil in Uganda ... You don't get that. It's rich ... Is it only in Uganda that it's red? Because in Kenya it is.

Bharti: But in Uganda it's even richer. I remember when it rained all the water, muddy water – we don't go out in that and walk in that ... And the smell you get when it rains ... I love that.

Manjula: We won't get that smell here!

Bhanu: I compare the soil [when gardening], the soil was so nice in Kenya [laughs]. It's really nice soft soil. Reddish sort of – here it's all clay ... It is hard work. Well, obviously, in Kenya, you don't do it all yourself. Mainly we have a gardener, but yes they sometimes feel it's so easy to grow fruits ... and the soil is very good. It's not difficult to plant trees, you know.

In the women's memories, the soil is remembered as fertile, almost magical, where papaya trees could not help but sprout, growing plants was easy, digging was easy, and gardening was not laborious. It is always the case that the soil in England cannot match up. The soil in England is hard work, and difficult. This is symbolic of their struggle to settle; their skin colour and bodies occluding an easy identification with Britishness.

Conclusions

In this chapter, I have mapped the routes of migration for South Asian women, which reflect the encountered landscapes and environments of a racialized post-colonial group living in the UK. Mapping these migratory geographies holds an important place in recognizing factors which influence landscape values of migratory populations in the UK. The connections with these geographies of migration are continually figured in daily life by the South Asian population, whose memories of landscapes constitute social narratives of heritage and cultural identity. Most importantly, memories of past landscapes have an impact on landscapes in Britain.

The British landscape is itself shifting in relation to this process of the presencing of these physical memories. The practice of gardening has become an expression of Asianness in England. This has materialized the biographical experiences and ecological knowledges of the women who have moved into the suburban scene. The landscaping of this scene is inscribed with echoes of ecological memories. These subtle expressions of migration through garden texture and aesthetics run counter to 'native' planting as expressed in contemporary writings about the preservation of a native planting aesthetic (Buchan, 1998; Thomas, 1998). These ecological paradigms of 'native' and 'non-native' species are politicized further in debates on ecological racism (Agyeman, 1990, 1991). For non-native gardeners, however, flowers are cultivated because they remind the grower of the 'other' place of being, but are not necessarily of that 'other' place. When South Asians migrated here, they came with their own imagined ecological portmanteau (Crosby, 1986). This reflected cultures of remembering and reconnecting with the soil, land and ecology of other nations. An ecological oral history is inscribed in the urban landscape and of suburbia, as part of reinscribing multisensory memories at home in England. These inscriptions include those national iconographies of identification and the intimate textures of the local left behind. These include the plant species, and the particular aesthetics of the biome of residence. This is a vernacular thread in the construction of new national cultures, after migration, layered upon the bigger discourses of nation and iconography of the nation. The placing of these textures in the geographical imagination and memory of diasporic groups can contribute greatly to understanding the process of making a home. Bright colours and familiar scents contribute to the cultivation of familiar landscapes on the domestic scale. The plantings of begonias, roses, jasmine, and coriander, in their different ways, are enfranchising environmental practices. They bind the women in an *embodied* way to an unalienated sense of being *in-place.* The gardens represent the women's new hybrid cultures. They store icons of the past, but also root the women firmly in the territory of Britain.

Social memories presented here are a means of routing diasporic groups, but also illustrate the continuing influence of landscape and intimate ecologies on identification with places. Migration (of plant and human species) disrupts essentialist notions of dwelling, belonging and *native species.* For migratory groups, however, experiences of nature are equally significant in their identification with local natures and national landscapes, as are their relationships with heritage narratives. As a result, new local ecologies emerge and are expressed through cultural practice as a process resettlement, and rerooting. They are material expressions of British Asian memory and cultures of enfranchisement. Body-memories are shards of *other* environments, evoked in the practices of everyday living, collaged together as a landscape of nostalgia, but also as a memory-history of Imperial landscapes, intimately experienced by post-colonial populations in the UK.

This relationship between past lived environments, imagined and idealized ones, and present lived landscapes has been termed a *triadic* relationship, pertinent to a post-colonial positioning theorized by Brah (1996) and Safran (1991). Materials of culture, such as organic materials in the home, resonate points of both geographic and historical identification which are significant in the affirmation of racialized identities in the British South Asian community. These materials of connection represent nodal points in a biographical journey, which, in turn, are symbolic of the political dynamic of making a home 'elsewhere'. These materials of culture are often situated as materials of negotiation of citizenship, belonging and national identity, in the process of 'centring' and 'positioning', after migration. They act as points of resistance to exclusive dominant cultures as well as offering points of engagement to an enfranchising idyll located in the past. These are memories which have materially shaped contemporary post-colonial domestic landscapes in Britain. Privileging these cultural materials and practices in geographical research offers researchers a means through which to examine the value of *landscape* itself to post-colonial communities living in Britain. Landscape memory, as embedded in domestic cultures, is presented here as an essential component of an attainable and inclusive *ecological* approach to landscape research with post-migratory communities. Reflected here are transcultural identities that connect belonging and being with particular ecologies. In terms of South Asian social history, what is needed is a *naturalization* of the archive; not only where ecological materials and memories are evoked as historical artefact, but where what is included in the archive is also stretched to incorporate *natural* organic materials of everyday life. The next chapter will take further this ecological approach to the understanding of historical memory.

Bibliography

Agyeman, J. (1991), 'The Multicultural City Ecosystem, Streetwise', *Magazine of Urban Studies* 7 (Summer), 21–24.

Agyeman, J. (1990), 'Black People in a White Landscape: Social and Environmental Justice', *Built Environment* 16: 3, 232–36.

Ahmed, S., Castaneda, C., Fortier, A.M. and Sheller, M. (eds) (2003), *Uprootings and Regroundings: Questions of Home and Migration* (Oxford: Berg).

Alexander, C.E. (2000), *The Asian Gang: Ethnicity, Identity, Masculinity* (Oxford: Berg).

Alexander, C.E. (1996), *The Art of Being Black: The Creation of Black British Youth Identities* (Oxford: Clarendon Press).

Anwar, M. (1985), *Pakistanis in Britain: A Sociological Study* (London: Edwin Packer and Johnson Ltd).

Anwar, M. (1979), *The Myth of Return: Pakistanis in Britain* (London: Heinemann).

Appadurai, A. (1997), *Modernity at Large: Cultural Dimensions of Globalization* (Minneapolis, MN and London: University of Minnesota Press).

Azaryahu, M. (2003), 'RePlacing Memory: The Reorientation of Buchenwald', *Cultural Geographies* 10: 1, 1–20.

Bhachu, P. (1985), *Twice Migrants: East African Sikh Settlers in Britain* (London: Tavistock).

Bialasiewicz, L. (2003), 'Another Europe: Remembering Habsburg Galicja', *Cultural Geographies* 10: 1, 21–44.

Blunt, A. (2003), 'Collective Memory and Productive Nostalgia: Anglo-Indian Homemaking at McCluskieganj', *Environment and Planning D: Society and Space* 21, 717–38.

Bohlin, A. (1998), 'The Politics of Locality: Memories of District Six in Cape Town', in N. Lovell (ed.), *Locality and Belonging* (London: Routledge).

Brah, A. (1996), *Cartographies of Diaspora: Contesting Identities* (London: Routledge).

Buchan, U. (1998), 'Blending In', *The Garden: Journal of the Royal Horticultural Society* 123, 596–98.

Butler, T. (ed.) (1989), *Memory, History, Culture and the Mind* (Oxford: Basil Blackwell).

Carrier, P. (2000), 'Places, Politics and the Archiving of Contemporary Memory in Pierre Nora's *Les lieux de mémoire*', in S. Radstone (ed.), *Memory and Methodology* (Oxford: Berg).

Carter, E., Donald, J. and Squires, J. (eds) (1993) *Space and Place: Theories of Identity and Location* (London: Lawrence and Wishart).

Castro-Gómez, S. (2002), 'The "Social" Sciences, Epistemic Violence, and the Problem of the "Invention of the Other"', *Nepantla: Views from South* 3: 2, 269–85.

Chambers, I. (1993), 'Narratives of Nationalism: Being "British"', in E. Carter, J. Donald, and J. Squires (eds), *Space and Place: Theories of Identity and Location* (London: Lawrence and Wishart).

Clifford, J. (1997), *Routes: Travel and Translation in the Late Twentieth Century* (Cambridge, MA and London: Harvard University Press).

Cohen, R. (1995), 'Rethinking "Babylon": Iconoclastic Conceptions of the Diasporic Experience', *New Community* 21: 1, 5–18.

Connerton, P. (1989), *How Societies Remember* (Cambridge: University of Cambridge Press).

Conradson, D. and McKay, D. (2007), 'Translocal Subjectivities: Mobility, Connection, Emotion', *Mobilities* 2: 2, 167–74.

Conway, M.A. (ed.) (1997a), *Recovered Memories and False Memories* (Oxford: Oxford University Press).

Conway, M.A. (1997b), 'Introduction: What Are Memories?', in M.A. Conway (ed.), *Recovered Memories and False Memories* (Oxford: Oxford University Press).

Conway, M.A. (1997c), 'Past and Present: Recovered Memories and False Memories', in M.A. Conway (ed.), *Recovered Memories and False Memories* (Oxford: Oxford University Press).

Crosby, A. (1986), *Ecological Imperialism: The Biological Expansion of Europe, 900–1900* (Cambridge: Cambridge University Press).

DuBois, W.E.B. (1994[1903]), *The Souls of Black Folk* (New York: Dover Publications).

Dwyer, C. and Bressey, C. (eds) (2008), *New Geographies of Race and Racism* (Aldershot: Ashgate).

Edgerton, S.H. (1995), 'Re-membering the Mother Tongue(s): Toni Morrison, Julie Dash and the Language of Pedagogy', *Cultural Studies* 9: 2, 338–63.

Fanon, F. (1967), *Black Skin, White Masks* (New York: Grove Press).

Fanon, F. (1961), *The Wretched of the Earth* (Harmondworth: Penguin).

Fanon, F. (1959), *Studies in a Dying Colonialism* (Harmondsworth: Penguin).

Ferguson, R., Gever, M., Minh-ha, T.T. and West, C. (eds) (1990), *Out There: Marginalisation and Contemporary Cultures* (Cambridge, MA: MIT Press).

Fortier, A.M. (2000), *Migrant Belongings: Memory, Space, Identity* (Oxford: Berg).

Gilroy, P. (2000), *Against Race: Imagining Political Culture Beyond the Color Line* (Cambridge: Harvard University Press).

Gilroy, P. (1994), '"After the Love Has Gone": Bio-politics and Etho-politics in the Black Public Sphere', *Public Culture* 7: 1, 49–76.

Gilroy, P. (1993a), *The Black Atlantic* (Cambridge, MA: Harvard University Press).

Gilroy, P. (1993b), *Small Acts: Thoughts on the Politics of Black Cultures* (London: Serpent's Tail).

Gilroy, P. (1987), *'There Ain't No Black in the Union Jack': The Cultural Politics of Race and Nation* (London: Routledge).

Gruneberg, M.M. and Morris, P.E. (eds) (1978), *Aspects of Memory* (London: Methuen).

Hall, S. (2000), 'Whose Heritage?: Unsettling "The Heritage", Re-imagining the Post-Nation', *Third Text* 49 (Winter), 3–13.

Hall, S. (1995), 'New Cultures for Old', in D. Massey and P. Jess (eds), *A Place in the World?* (Milton Keynes and Oxford: Open University Press and Oxford University Press).

Hall, S. (1990), 'Cultural Identity and Diaspora', in J. Rutherford (ed.), *Identity: Community, Culture, Difference* (London: Lawrence and Wishart).

Hopkins, P. (2007), 'Young People, Masculinities, Religion and Race: New Social Geographies', *Progress in Human Geography* 31, 163–77.

Hurnong, A. and Ruhe, E. (eds) (1998), *Postcolonialism and Autobiography* (Amsterdam: Editions Rudopi).

Jazeel, T. (2003), 'Unpicking Sri Lankan "Island-ness" in Romesh Gunesekera's Reef', *Journal of Historical Geography* 29: 4, 582–98.

Kaur, R. and Kalra, V.S. (1996), 'New Paths for South Asian Identity and Musical Creativity', in S. Sharma, J. Hutnyk and A. Sharma (eds), *Dis-Orienting Rhythms* (London: Zed Books).

Knott, K. (2005) 'Towards a History and Politics of Diasporas and Migration: A Grounded Spatial Approach', presented at 'Flows and Spaces', Annual Conference of the Royal Geographical Society/Institute of British Geographers, Diasporas, Migration and Identities Working Paper 1, http://www.diasporas.ac.uk.

Kritzman, L.D. (ed.) (1996), *Realms of Memory: Rethinking the French Past*, vol. 3 (New York: Columbia University Press).

Legg, S. (2005), 'Contesting and Surviving Memory: Space, Nation, and Nostalgia in *Les lieux de mémoire*', *Environment and Planning D: Society and Space* 23, 481–504.

Lim, J. (2008), 'Encountering South Asian Masculinity Through the Event', in C. Dwyer and C. Bressey (eds), *New Geographies of Race and Racism* (Aldershot: Ashgate), 223–38.

Lovell, N. (1998), *Locality and Belonging* (London: Routledge).

Lowenthal, D. (1985), *The Past Is a Foreign Country* (Cambridge: Cambridge University Press).

Lowenthal, D. (1979) 'Environmental Perception: Preserving the Past', *Progress in Human Geography* 3: 549–59.

Lowenthal, D. (1975), 'Past Time, Present Place: Landscape and Memory', *Geographical Review* 64: 1, 1–36.

Lowenthal, D. and Prince, H. (1965), 'The English Landscape', *Geographical Review* 3, 319–46.

MacNaughten, P. and Urry, J. (1998), *Contested Natures* (London: Sage).

Malik, S. (1992), 'Colours of the Countryside – A Whiter Shade of Pale', *ECOS* 13: 4, 33–40.

Massey, D. and Jess, P. (eds) (1995), *A Place in the World?* (Milton Keynes and Oxford: Open University Press and Oxford University Press).

McLaren, P. and Pinkney-Pastrana, J. (2000), 'The Search for the Complicit Native: Epistemic Violence, Historical Amnesia, and the Anthropologist as Ideologue of Empire', *Qualititative Studies in Education* 13: 2, 163–84.

Menchu, R. (1984), *I, Rogoberta Menchú, an Indian Woman in Guatamala* (London: Verso).

Modood, T. (1994), 'Political Blackness and British Asians', *Sociology* 28: 4, 859–76.

Modood, T. (1992), *Not Easy Being British* (London: Runnymede Trust).

Modood, T. and Werbner, P. (eds) (1997), *The Politics of Multiculturalism in the New Europe: Racism, Identity and Community* (London: Zed Books).

Mohammad, R. (2007), '*Phir Bhi Dil Hai Hindustani* (Yet the Heart Remains Indian): Bollywood Cinema, the "Homeland" Nation-State and the Diaspora', *Environment and Planning D: Society and Space* 25, 1015–40.

Mohammad, R. (2001), '"Insiders" and/or "Outsiders": Positionality, Theory and Praxis', in M. Limb and C. Dwyer (eds), *Qualitative Methodologies for Geographers* (London : Arnold), 101–20.

Morris, P. and Gruneberg, M. (eds) (1978), *Theoretical Aspects of Memory* (London: Routledge).

Morrison, T. (1990), 'The Site of Memory', in R. Ferguson, M. Gever, T.T. Minh-ha and C. West (eds), *Out There: Marginalisation and Contemporary Cultures* (Cambridge, MA: MIT Press).

Nayak, A. (2006), 'After Race: Ethnography, Race and Post-Race Theory', *Ethnic and Racial Studies* 29, 411–30.

Nora, P. (1996), 'Era of Commemoration', in L.D. Kritzman (ed.), *Realms of Memory: Rethinking the French Past*, vol. 3 (New York: Columbia University Press).

Nora, P. (1989), 'Between Memory and History: *Les lieux de mémoire*', *Représentations* 26, 7–25.

Nora, P. (1986), *Les lieux de mémoire*, vol. 2: *La Nation* (Paris: Gallimard).

Nora, P. (1984), *Les lieux de mémoire*, vol. 1: *La République* (Paris: Gallimard).

Phillips, D. (2006), 'Parallel Lives? Challenging Discourses of Self-Segregation in the British Muslim Population', *Environment and Planning D: Society and Space* 24, 24–40.

Puwar, N. and Raghuram, P. (2003), *South Asian Women in the Diaspora* (Oxford: Berg).

Radstone, S. (ed.) (2000), *Memory and Methodology* (Oxford: Berg).

Ramji, H. (2006), 'Journeys of Difference: The Use of Migratory Narratives Among British Hindu Gujaratis', *Ethnic and Racial Studies* 29: 4, 702–24.

Relph, E. (1976), *Place and Placelessness* (London: Pion).

Rutherford, J. (ed.) (1990), *Identity: Community, Culture, Difference* (London: Lawrence and Wishart).

Safran, W. (1991), 'Diasporas in Modern Societies: Myths of Homeland and Return', *Diaspora* (Spring), 83–99.

Said, E.W. (1978), *Orientalism* (Harmondsworth: Penguin).

Samuel, R. (1994), *Theatres of Memory* (London: Verso).

Sayyid, B.S. (1997), *A Fundamental Fear: Eurocentrism and the Emergence of Islamism* (London: Zed).

Sharma, S., Hutnyk, J. and Sharma, A. (eds) (1996), *Dis-Orienting Rhythms* (London: Zed Books).

Spivak, G. (1998), 'Three Women's Texts and Confessions', in A. Hurnong and E. Ruhe (eds), *Postcolonialism and Autobiography* (Amsterdam: Editions Rudopi).

Spivak, G. (1993), *Outside in the Teaching Machine* (New York and London: Routledge).

Stoll, D. (1999), *Rigoberta Menchu and the Story of All Poor Guatemalans* (Boulder, CO: Westview Press).

Thomas, C. (1998), 'Community Centre', *Gardeners' World*, 50–53.

Tolia-Kelly, D.P. (2004), 'Processes of Identification: Precipitates of Re-memory in the South Asian Home', *Transactions of the Institute of British Geographers* 29, 314–29.

Tolia-Kelly, D.P. (2002), 'Iconographies of Diaspora: Refracted Landscapes and Textures of Memory of South Asian Women in London', PhD thesis, Department of Geography, University College London.

Tolia-Kelly, D.P. (2001), 'Iconographies of Identity: Visual Cultures of the Everyday in the South Asian Diaspora', *Visual Culture in Britain* 2, 49–67.

Turner, B. (2002), 'Cosmopolitan Virtue, Globalization and Patriotism', *Theory, Culture and Society* 4: 19, 45–63.

Urry, J. (1990), *The Tourist Gaze: Leisure and Travel in Contemporary Societies* (London: Sage).

Vertovec, S. (1999), 'Conceiving and Researching Transnationalism', *Ethnic and Racial Studies* 22: 2, 447–61.

Young, R.J.C. (2008), *The Idea of English Ethnicity* (Oxford: Blackwell).

Chapter 5

Material Memories: Visual and Material Cultures in the South Asian Home

Introduction

The aim of this chapter is to illustrate the ways in which social memory and systems of identification with landscape, for post-colonial migrants, are integral to the everyday environments of home. Home in this research is the site of historical artefact, social history and connectivity with a race-history that is core to the South Asian diaspora in London. Visual and material cultures are material to the diaspora; they at once *locate* the diaspora, historically, geographically and culturally. Also these cultures refract past locations and identifications; they harbour the heritage of the diaspora's *locatedness*. This chapter will start by working through a conceptualization of *re-memory* and will situate ecological memories. Ecologies of living and belonging are signified through these forms. The chapter ends with a focus on the material nature of the visual and considers the interwovenness of the formal registers of visuality and materiality.

Paul Connerton's *How Societies Remember* (1989, p. 37) accedes to Halbwachs' (1992) claim that 'the idea of an individual memory, absolutely separate from a social memory, is an abstraction almost devoid of meaning.' In the groups that I have worked with, there is a simultaneous synthesis between understandings of individual identification with a heritage of being 'South Asian' and collective identification despite the heterogeneity of migration routes and birthplaces. One strand of connectivity that gels this identification is the ways in which places, sites and situations are recalled in group discussions; memories of situatedness are shared and consolidate the group's identification with a past, journey, society and heritage. The question that is live for both Connerton and Halbwachs is, how do collective memories that form social histories get passed on to the next generation? In this instance, Connerton turns to Marc Bloch (1886–1944), a French Marxist anthropologist. Bloch (1954) argued that oral narratives, as part of the socialization of the young, are the source of learning, education and consolidation of social identity through memories and recollections in group narratives. These acts of 'transfer' are the social glue that retains the collective without denuding or negating changes. For Halbwachs, Bloch and Connerton, dynamism within this process is essential for the narratives to retain credibility, and to do that they need to be flexible. These stretching, shifting and 'open' narratives enable a social memory to be retained despite mobility, shifts in social organization and re-territorialization. In my research, the group identification with a past 'territory of culture' is self-determined. This is a collage of places, landscapes, experiences of oppression and joy that are recognizable as being part of a South Asian experience of diaspora, post-colony. The disparate biographies do not deter a self-determined account of Asian collectivity. The aim here is not to collapse the differences, but to privilege the *social* nature of the women's memories. To retain a social sense of memory, to materialize the 'transfer' that Bloch recognizes, amongst the South Asian diaspora, I have focused on the tangible materials of visual and material cultures in the home. In Williams' (1979) account, the cultural materials of a society garner a society's vision of itself. William focused on the novel as a cultural material in circulation. To enable research on mobile communities who have a limited canon of literature in circulation, I have focused on the material cultures in the home. These material objects revealed much about social memory. In this chapter, I will work through two forms of memory; ecological memory and re-memory; embedded in these are ecological memories that connect the women with their shared 'territories of culture'.

Remembering *Home* at Home: Materializing South Asian Re-Memory

Material cultures, through their installation, are critical in the formation of new political identities, carving out new landscapes of belonging. The 'home' is the stage for emitting history (Samuel, 1994). These solid precipitates are where memories of past accounts accrue. Signification of identity, history and heritage, through these material cultures, relies on the continuing dependence on the past for sustenance in the present. These material cultures secrete an essence of security and stability. Ironically, these material foundations are sometimes transient, ephemeral things, which fade, tear, fragment, dissolve and break. Individual objects relate to individual biographies, but are simultaneously significant in stories of identity on national scales of citizenship, and the intimate domestic scene left behind. The new site of home becomes the site of historical identification, and the materials of the domestic sphere are the points of signification of enfranchisement with landscapes of belonging, tradition and self-identity.

After mobility, these new contexts for material artefacts refigure the narration of the past imbued within them. Memory is an important political tool, grounding both individual memory and collective cultural heritage stories. These processes are not exclusive to the South Asian population; in fact, other writers have looked at different migrant communities and their valuing of domestic artefacts as stores of cultural narratives and memorialized biographical narratives (Boym, 1998; Lambert, 2001). The presence of these materials of heritage disturbs and shifts notions of Britishness. If we look at the collage of material cultures in the British Asian home as layered with aspects of memory, they become historical inscriptions within the domestic landscape. Material cultures are critical in relation to the new sites of identity–territory relations; memory-history, as I have posited it, is activated *in relation* to the new context of living. These domestic inscriptions record the post-colonial positioning that informs a politics of South Asianness within a multicultural landscape. Imbued within this political orientation is a geography of being, belonging and making home, linked directly with a post-colonial history. Within this analysis, 'multiple provenances' emerge (Parkin, 1999, p. 309), where the notions of 'home' and 'origin' are not fixed in one locus. Memory-history counters the unbounded notions of 'Asian' ethnicity (biological) and nationality (cultural) through a system of collective logic that is a collectively remembered and valued memory-history (Tolia-Kelly, 2002).

Re-memory is figured in this research as memory that is inscribed with a race-politics that is part of everyday social discourse (Brah, 1999), within the South Asian diaspora. As Leela Gandhi (1998) has notably expressed, it is difficult to consider the post-colonial position as being truly *post-*, as the struggles that these societies face are intrinsically linked to social, cultural, economic and political structures which are controlled and shaped by those *ex*-colonial powers. Re-memory is memory that is encountered in everyday life, but is not always a recall or reflection of actual experience. It is separate from memories that are stored as site-specific signs linked to experienced events. *Re-memory* can be the memories of others as told to you by parents, friends, and absorbed through day-to-day living. These re-memories contribute to a sense of self beyond the linear narrative of events, encounters and biographical experiences that usually forms our sense of family social history. Re-memory is an inscription of a body or event in time in place, which is touched, accessed, or mediated through sensory stimuli. A scent, sound or sight can metonymically transport you to a place where you have never been, but which is recalled through the inscription left in the imagination, lodged there by others' narratives. This form of social geographical coordinate is not always directly experienced but operates as a significant connective force. Re-memory is a resource for the sustenance of a sense of self that temporally connects to social heritage, genealogy, and acts as a resource for identification with place. The origin of the concept of re-memory is Toni Morrison's novel *Beloved* (1987, pp. 35–36):

> I was talking about time. It's so hard for me to believe in it. Some things go. Pass on. Some things just stay. I used to think it was my rememory. You know. Some things you forget. Other things you never do. But it's not. Places, places are still there. If a house burns down, it's gone. But the place – the picture of it – stays, and not in just my rememory, but out there in the world. Someday you be walking down the road and you hear something or see something going on. So clear. And you think it's you thinking it up. A thought picture. But no. It's when you bump into a rememory that belongs to someone else. The picture is still there and what's more, if you go there – you never was there – if you go there and stand in the place where it was, it will happen again; it will be there for you, waiting for you.

Morrison proposes re-memory as a form of race-memory. Through bodily encounters in space, Morrison promotes the idea that you can bump into the oppressive experience of slavery and lost narratives. For Morrison, her writing through of re-memory narratives makes the Black experience and the collectively held record of slavery into text. The text makes material this intangible slave history. Re-memory inscribes the history of Black oppression within the public domain, as a resource for the centring of self through the past, for the Black community. Re-memory mediates between now and the separation of slaves from their families, homes, and lives. Because re-memories are *socially* experienced they become material, and are shared amongst the African-American community. These are narratives beyond the traditional form of oral testimony; these operate in public space, and have, for the individual encountering them, an intimate resonance with past narratives of others not known. Their effect is constitutive; the memories form the social memory of the community, effectively linking the group through a sense of collective history. Re-memory's reinscription in the environment is a political force, countering exclusionary heritage discourses, and providing a form of synthesized embodied heritage. For those other diasporas that have experienced post-colonial labour migration, there is also a sense of rupture with a heritage story of a community, not recorded formally. My aim is to trace the social memories at the site of home where memories traverse, are stored, exchanged, encountered and materialized.

'Home' has many incarnations within social memory of the diaspora. Intertwined in the social memory of the group are imagined and lived landscapes of home, and utopian senses of 'home' where 'home' continues to be a source of identification. Home, as feminists have argued, is not always a refuge; it continues to be a site of gendered oppression and violence. For Black women, home is the landscape used strategically as a site for refuge from the alienating experience of racism and marginality. Within my research, however, home is a conceptual tool to discover different means of engagement with lived landscapes; home harbours cultures of 'making home' and enfranchisement to new places and citizenships. Possessions operate as material nodes that symbolize, refract and resonate with the diasporic journey; they are connective markers to geographical nodes of identification. Through their prismatic nature, 'other' lives, lands and homes are made part of this one. Material possessions operate as a buffer against exclusive national cultures by encompassing a collage of familiar textures. These 'home possessions' (Miller, 2001) presence the social memories; they constitute precipitates of narrated histories, and artefacts of heritage and tradition. These are souvenirs from the traversed landscapes of the journey, as well as signifiers of 'other' narrations of the past not directly experienced; these narrations are termed 're-memories' (Morrison, 1987, 1990). These are identified as precipitates of lived racial politics in the form of material artefacts (Csikszentmihalyi and Rochberg-Halton, 1981; Mehta and Belk, 1991).

Enshrining Heritage – Mandirs, Medina and Religious Iconography

Firstly, I focus on the sacred artefacts, in the form of shrines and religious iconology within the homes of the women interviewed. In all the women's homes, there was a significant religious space. The mandir is the sacred home temple that each of the Hindu women had (Figures 5.1–5.3). These shrines/artefacts operate as reminders of moral values, symbolize stories in the *Bhagavad-Gita*[1] and the *Ramayana*,[2] and offer a collection of religious icons which reflect different life values – wealth, prosperity, health and righteousness. My interest was in the materiality of these objects in relation to the women's feelings of home, belonging and identity. It is quite clear from the tone and feeling with which the women describe the religious icons and objects how much their lives are intertwined with the existence of the shrine itself and the variety of icons and other materials held within:

> **Hansa:** Previously in my old house I just had a cabinet which I made – I didn't have a mandir or anything. Recently in the last three–four years, in Bombay I managed to get a little mandir, 4 feet by 4 feet. It's just a wooden mandir, really tiny. That's because there were problems of bringing it down here. I had made my mandir out of a small cabinet, furniture, not exactly a mandir shape or anything. We had this before. Murtis are in there too. Slowly, slowly I've collected them.

These Hindu mandirs individually are dynamic sites whose content, size and aesthetics shift over time. They usually start as a small place in the house, such as a corner that will not be defiled during menstruation, and so they are often in cabinets, cupboards or entire rooms that can be locked up. The shrines hold religious relics such as *gangajal*, water from the Ganges; *vibhuti*, the sacred dust from incense burning at pilgrim sites in India or at other temples; and *chunis*, small pieces of embroidered cloth used on statue murtis (icons) from temples. These are blessed during *aartis* – special daily prayers. Murtis themselves have been purchased at sites of pilgrimage, Asian cloth shops, grocers, or even sweet shops. Alongside these are chromolithographs (Pinney, 1995) – vividly coloured, paper images of icons, gurus and saints – and other images which are made by combining the processes of photography and lithography with layers of colour printed on a photograph.

The presence of these shrines significantly contributes to heritage-practices and a sense of cultural nationalism. The images in the shrines are iconographic. These significations are locked into a connection with place; therefore, they are iconographic as both a religious sign and a sign of a particular landscape. They represent an inter-ocular field (Appadurai and Beckenridge, 1992). Essentially, this means that the texts can be read simultaneously as singular texts, or as icons with

---

1 This is the sacred Hindu text which is part of the *Mahabharata*.
2 The *Ramayana* is the Sanskrit epic of Rama, an incarnation of Vishnu.

multiple significations. The meaning of an individual religious image can operate as a sign for that religion, a sign for the story of an individual icon, a signification of India (through its visual grammar), or a personalized metonymical symbol of the moment that it was received (from a parent or a priest), or a particular space that it is connected with (site of pilgrimage). Through these characteristics, any individual image oscillates between encompassing religious, spatial, and historical iconography (Pinney, 1997, p. 111). These signs are part of a collective, visual vocabulary for the South Asian community; social connections with a sense of communal identity are made through visual registers of colour, texture, sound and scent. The *gangajal* is a container (usually copper) holding water from the Ganges. On sight, in the shrine, it operates as a symbol of the sacred site of the Ganges and of its regenerative powers. It is also a symbol of the nation, of India. The Ganges is the site of pilgrimage and therefore the *gangajal* acts as a souvenir of that place, and the moment of visiting; it may also be a metonym for others' narratives about the Ganges, and for Hinduism and the regenerative power of Ganges water. The *gangajal* makes tangible these various narratives and tales, and consolidates various senses of place linked to the object; they are biographical and memorialized connections which are used to figure a life in the UK, post-migration.

These artefacts are *possessions*; they are remnants, the physical debris of the social imagination that is re-memory. The shrines are significant in their value as points of engagement with being, living and developing a history in England. Shrines are dynamic; they allow the growth of a collection of sacred and blessed pieces, but they are not limited to religious sacred objects and icons. They are considered to be at the heart of family relations. Sometimes they are sites where the family genealogy can be traced. The shrines become places where, due to their sacredness, important family objects are placed. The shrines incorporate family photos of those who have passed away – grandfathers, great-aunts, grandmothers, and great-uncles. The shrine becomes collaged, continually superimposed with objects reflecting intimate and sacred life moments which are preserved and treasured; such items might include a rosary given by a father, a piece of gold received on a wedding day, seashells or *shivlings* (representations of the Hindu deity Shiva) that were in the family home prior to marriage. Such items are all composed and united in the laying out of the shrine. This process of collage is an accrual of the sacred, and is emotionally valued by the family. Adding objects to the shrine is inscription. Sometimes families place letters at the shrine – job applications and offers, letters of award and even travel tickets, so that all these possible journeys can be blessed. Symbolic representations of the first breath of a baby, the first job, the wedding, and all the rites of passage and major life stages can sometimes be traced. By the *murti's* presence, the women believe measures have been taken to prevent obstacles, mishaps and general misfortune. Intensely personal prayers are recited here – celebrating, requesting and proffering. These are a set of intense moments and small things, miniature icons representing larger things, moments, and connections. Embroidered onto these textures of the object is a set of relationships between biographical and national and/or cultural identifications. The shrines accrue and secrete meanings connected to remembering the experiences and narratives of others. Over time the shrine accumulates layers of meaning. It is believed to emanate protective vibrations. It is a spiritual place before it is a purely religious site. Its significance grows with time along with its representativeness of family biography; the objects form a significant collage of embedded events, moments, and aesthetic imprints. These private, personal moments are inscribed physically. Their meanings shift in context, as the shrine is moved when the family move and thus traces migration at every scale. The mandir is the responsibility of the matriarch. Below Hansa describes her memory of her mother and her mother-in-law, who had different religious beliefs and commitment and the memory of their lived landscapes:

> **Hansa:** In India it [public temple] was opposite our home – always wherever you go, it was so much on your way that you could quickly pray and leave. For my mum it was very important despite being partially sighted. She never used to go out too much. But she used to make her way up to the Derasar [Jain temple] … It's not in an open area, but residential area. It's nice, Derasar is always nice, cool places … India is very hot but these places are very cool. They hold and bring a sense of inner peace … It [the mandir at home] brings me peace. It's basically important for what I do and say … I apologise for what I've done … Without it, I would still be able to focus, but my boys would not have seen it, and they would not be used to us seeing it. I can still close my eyes even if I don't have a mandir, I can visualize. Because I have seen day in and day out that Derasar there. But J. and the boys they have not seen it.

The mandirs refract collages of social and spiritual life. Their presence also signifies cultural, social and moral discourses and practices. Therefore the shrines reflect back into the home many moments and spaces. They both absorb daily activity and secrete a historical context to these. As they trigger memories, they operate on many scales of time, their reflectivity enhances the meaning of these shrines personally, culturally, and they in turn produce new meanings and memories for the women and their families. For Hansa's family, the shrine becomes a source of re-memory of Hansa's experiences of these religious landscapes of the past (in various life stages). The shrine makes tangible her stories and others' stories told through her narratives. Various places and moments are brought into the home through these material links. Here memories are made present through the matter of the shrine.

The relationship that Lalita describes below illustrates the interpolation of public and private religious practices into a critical cultural expression. These artefacts embedded within the social and cultural life of India form her sense of self through their presence:

Figure 5.1   Lalita's mandir

Figure 5.2   Hansa's mandir

Figure 5.3   Shazia's mandir

Lalita: I think religion is part of my culture. Especially being brought up in Delhi, for me it was not two different things. I just couldn't say this is my culture. OK, it's Kathak, and classical music, and touching elders' feet, respecting them as my culture. And this is my religion. No, it was a fusion of both, it was blended together. So it was like … all festivals were not only religion, not only culture but fun also. So it was like a blend of everything. And we enjoy it. Like sometimes we were 40 of us going together. My father had eight brothers, and sisters, their spouses, their children, and cousins. Everybody going, booked like four coupes of the train, and rent a bus at Jamnu.

Lalita describes the difficulty of defining her culture without religion. But the religious excursions she describes are social events. They have codes, rites and rituals of their own. Religion is not bounded within strict definitions of personal relationship with God, or limited to certain religious rituals. The religious learning is part of socialization in the home. Each daily task has a religious moral code inscribed. For example, 'eating' has rules and codes. Defining the moral order of family hierarchy was part of the function of religious excursions. However, the religious text was not privileged over the social function of the trip. Religious texts and iconographies are woven through daily practices and rites. To some degree the shrines operate in the same sphere; their presence socializes family members into religious teachings and their influence on daily life. The shrine itself, or objects and icons within, are incorporated into the events of different life stages. The shrine activates a connection biographically and spiritually. The intensely personal spheres are shot through with religious moral codes and practices, which are also inscribed onto the shrine. This is done through adding objects, as well as inscribing contemporary meanings onto the materials within the shrine.

Shazia: This is my mandir, it's the first room in the house. It's got our papers, the piano … This room has always been a mixture – everything has got some meaning. You know like any important letters we have. … Everything has over the years been picked up. Like this is from my mum's place. This one I got from her mandir … the mandir has photos of Lalit's parents and my parents.

The shrine's aesthetics and icons invite spiritual focus; they instil a sense of cultural education and integrity. Shazia believes that her family are lost without this reference point. Through gazing at the icons, they are required to situate themselves in relation to the heritage practices of grandparents, parents and their kinship networks. For the diaspora, a *positioning* (Hall, 1990) is attainable through its presence. This is a complex positioning for the South Asian diaspora which can at different moments define itself in relation to both religious identity (Hindu, Islamic, Jain, Sikh or Christian) and identity as post-colonial Indian nationals. My focus here is on the shrines' connective power to other landscapes rather than on religious ideology. The shrine is symbolic of a cultural identity linked to a space pre-migration, and its textures are central to this recalling process. The process of recall does not connect necessarily to lived and experienced memories or events; these collage with narratives of 'others'. The scent, touch, sight and sound of prayer; the prayer bell; the scent of camphor, incense, and sandalwood; the feel of *vibhuti*, cloth, ghee and cotton wool; and the sight of icons, of vermilion, are real textures, referencing religious rites and practice; they are all part of situating and socializing children, centring them in relation to heritage stories and the social significance of these texts and textures in them. Importantly, the shrines enable and assert the practices of being Asian in relation to *other*ness. The shrines represent an iconography of cultural integrity, but their dynamism allows for greater relevance to everyday life.

The Hindu shrines trigger re-memories of sites of spiritual well-being. Shilpa describes the intimacy she solicits through her shrine with the temples in Mumbai. This labyrinthine city has alleyways leading off main streets with wayside shrines which are points of congregation for Hindus. These shrines contribute to a landscape of righteous, moral living which extends to Shilpa's feelings of belonging to a dynamic and connected community. In her description, the alleyways have shrines, which are open day and night; they are lit and visited. When recalled, they are points which offer meaning and light beyond the alienation experienced in the UK:

> **Shilpa:** Everywhere [in India] is spiritual. In alleyways there are shrines. Your soul gains peace, you know? Here, [in England] travelling to the temple is very difficult. … In India you get to see people. Any place, you get to offer prayer. On the streets, roads you come across temples small and large.

This religious landscape extends throughout India, and these shrines are duplicated in the home, as miniature versions. But, of course, being situated in the home means that they exclude the social networking that public shrines allow. They allow connection only to the grander religious narratives of nation and Hinduism, but exclude the daily social and psychological enfranchisement that the existence of places of congregation offers. Public celebrations and festivals also take place in the everyday streets and roadways. They are not limited to single temples or focal points. The crossroads of streets are where pedestrians join the celebrations; there is music, food and religious songs and prayer. For Shilpa, it is not the event that is important but the fact that the streets and alleyways are all made part of the inclusionary landscape around. She does not feel alienated or distant. The religious activity ensures an intimacy, and a sense of belonging. This is translated also into the *look* of the streets and commercial districts. Religious icons are in all the shops and shrines, and there is a smell of incense and camphor and an aesthetic which she has tried to replicate in her own temple at home. The icons, the music and the prayers through which she celebrates them are at the heart of her home life.

Re-memory is a significant expression of double-consciousness for racialized groups (DuBois, 1994[1903]), experienced in the everyday and signified through expressive cultures (Gilroy, 1993a). For post-colonial people, there exists a consciousness that is fractured, shifting between identification with a perceived utopian, pre-colonial identity and one that is shaped by an imposed colonial regime of race-definition, and the lived experience of being a post-colonial within Britain. It is within the arena of expressive cultures that double-consciousness has been made evident. Black artists, writers and musicians have engaged with the 'doubleness' of being within the heart of the West, but figured as 'Other' within that nation's history and society (Araeen, 1991, 1992a; hooks, 1994, 1995). The history and heritage of Black slavery and race oppression informs a consciousness – a historically located consciousness, referenced through contemporary expressive cultures, and which is expressive of the race-positioning of contemporary Black identities. Morrison's re-memory is a haunting of the entire Black race by the inhuman experience of slavery (Finney, 1998); 'to pursue a future without remembering the past has its own deeper despair' (Horovitz, 1998, p. 97). The experience of the past informs the race-positioning of all post-colonial migrants in the present, and the process of the activated relationship depends on the needs of the *tellers* and the *listeners* (Sale, 1998, p. 3). The call and response aspect of this racial positioning fulfils the function of informing and anchoring current identity processes with an historical memory-history that references *tradition*, *heritage* and *history*. Re-memory is a process engaged with the 'interiorlife' of post-colonial groups who are constantly negotiating between past landscapes and the present territories of citizenship. By 'interiorlife', I refer to the internalized experience of rupture, alienation and non-identity during the period of post-migration. Re-memory is imbued with the sentiments of loss and absence, but it is critical in the politics of identification, which is so pertinent in contemporary global politics of migration, race, and heritage. The struggles of dealing with racism, fascism, and the social and economic landscape of a Black race-identity are present within the pain motif of re-memory. This reflects the 'interior life' of the experience of diasporic migration. Re-memory, being a site for the sustenance of the self, regenerates cultural and social collage of events; it embodies a set of physical, emotional and geographical coordinates from which to forge a cultural heritage.

The intensity of re-memory's power operates as a catalyst between Black communities past and present; its presence actively shapes the tones and forms of cultural expression through language, music, text and dance. The pain motif in Gilroy's and Morrison's work is complex and yet singularly defined by the experience of the slave, bodily and spiritually subjugated to the slave owner's designs and the colonial regimes of signification and value (Fanon, 1959, 1961, 1967). Within this pain motif is the experience of rupture from family, land and biological and cultural identity. The significance of this pain motif for other post-colonial groups has yet to be fully researched, and it would be wrong to cite the painmotif as equally manifest amongst all post-colonial peoples, including South Asians living in Britain. The regimes of truth and order were significantly too varied and complex in East Africa and India to be discussed in this way. However, the experience of rupture, loss, and vulnerability in social, cultural and biological terms has been a factor in diasporic identification. There are commonalities in their positioning today as post-colonial subjects, where the experience of loss and rupture is translatable to the experience of British Asians. However, the pain motif of enslavement is demonstrably unique and particular to the African population of enslaved men and women, who appear anonymously in history but are significant in the formation of individual race-histories today. In this chapter, the pain motif is a transitory feature of South Asian re-memory; however, traces exist within the text and the voices of the participants, alongside the creative, positive, and affirming qualities of remembering.

Within this analysis, I do not wish to collapse slave memory with the oppression of the South Asian – this would be an injustice. This is not an attempt to transpose re-memory to another ethnic group, but to consider the power of re-memory in the South Asian 'processes of identification', which are also post-colonial and diasporic. I wish to recover the links between the Asian experience of post-colonial positioning and the African-American one, through the political use of 'BlackBritishness'. There are simultaneously specific and different formations of *Blackness*, which commonly have racism as a driving force of their constitution. 'Asianness' is not simply a parallel post-colonial diaspora to the Black African diaspora; it has different colonial histories and has resulted in various post-colonial geographies. The political definition of Black has expressed the ways that these solidarities have been theorized and played out. British Asian expressive cultures have been the site of political struggles and have made a critical contribution to identity discourse. As Kobena Mercer (1990a, 1990b), writing about the 1970s and 1980s, states, Asian expressive cultures have been critical in the movement for race equality and justice. In the 1970s, voices such as Amrit Wilson (1978) and Pratibha Parmar were significant examples of political activists who were culturally Asian and politically Black. Politically, the struggle against racism was considered a universal responsibility. In this practice, therefore, the Asian activists were not a distinctive group with separate interests. However, since the decline of the political climate of the early 1970s, writers such as Tariq Modood (1990, 1994) have favoured a culturally essentialist approach. Modood privileges religious cultural identity over (firstly) the shared experience of Asians of racial oppression, and (secondly) the employment of 'Black' as a political banner raised in opposition to divisive cultural politics. Modood diagnoses an oppression of Asians as mediated through the politics of religion, as has been evidenced in geopolitical rhetoric on a global scale, since 9/11.

Ecological Portmanteau

In this section, I want to contextualize the effects of the sacred artefacts with rather different cultures within the home. These refract a different set of visualized memories, but operate similarly within the regime of a social history of the South Asian group. Landscapes that are represented or refracted through crafts and curios more crudely represent lived environments of the past which have become icons themselves. The touristic curios and craft objects that the women in the groups gather in the home have varied biographies; their routes are not exactly the same as those of their owners and are often part of the commodity culture. They are bought as 'souvenirs' or deliberately connecting devices *after* migration. They are situated as symbolic of other lives, narratives and national identifications. For the women from East Africa, animal products and dark wood sculptures feature quite heavily in their homes and in their stories of East Africa; copper plates, ivory products and Masai Mara curios were all seen as valuable in their presence in the women's homes in England. All of these things hold a *cultural vitality* (Gell, 1986, p. 114). They are not always deemed as objects with functions, but their meanings are infused with the biography or socio-cultural markers of the owner. When the women were living in East Africa, these curios were part of the commodity markets; they are defined differently from the British context. They are imbued with a sense of Africanness, but were sold as tourist souvenirs in a mass market of similar tokens. The women in the groups made clear that (when they were living in East Africa) these things were considered kitsch or lacking in style. They were tacky remnants of a commodified ecosystem – elephant tusks, zebra skin, ivory necklaces, and leopard skin handbags are all examples of the slicing of the African savannah into saleable touristic souvenirs. Within the British context, however, these items bear the prismatic qualities of the shrines, refracting memories of other lived landscapes. As Hirsch and O'Hanlon (1995, p. 23) recognize:

> [t]here is not one absolute landscape here, but a series of related, contradictory moments – perspectives – which cohere in what can be recognised as landscape as a cultural process.

> **Bhanu:** Like flamingo feathers with some ornaments on their table.

> **Shazia:** Or those Mombassa gates in that copper, that is a must in most Kenyan houses.

> **Shanta:** Nobody used to keep it in their home. We didn't have anything even on the walls, no. Nothing in the house, only after we left I think we are thinking of. … My husband collects these David Shepherd pictures. These are mostly animal scenes, elephants, and the African savannah feature largely in all of them.

Objects bought after leaving East Africa accrue meaning in their new context. In the process of being replaced, and replanted in the UK, these touristic curios are imbued with different values and meanings. What occurs is a reconstruction of value (Geary, 1986). Whilst the women were living in East Africa, the objects were just background things; together they formed a collage, as part of everyday life, a backdrop to African life. The diasporic journey imbued them with a heightened significance. After the move, they are created anew, in the process of their circulation. They may have been bought as objects of little value, but in the

Figure 5.4   Batul's African copper engraving

Figure 5.5   Shanta's African copper engraving

context of migration they are bought, given as gifts, or ordered at great expense to ensure their contribution to the cultural landscape of the home in England. Their intrinsic value is limited, but their symbolic value shifts through time; their contexts have been reconstructed. The activation of these signs and symbols occurs through the processes of remembering this 'other' landscape through them. Their earlier value is heightened in their being dislocated along with the owner, but in a social context of recognition and signification. Thus, an elephant tusk, or a zebra skin bag, not only is valuable to the owner personally, but also has a role in signifying to others who see it in the home; it has a biographical and cultural vitality which inevitably resonates with others within the community and without. Such items resonate with biographies, and the narratives of others that inform a store of heritage values; the owners' re-memories are contained, refracted, and mediated through them. These curios become treasures; they are revered through their reconstruction and in turn reconstruct their contexts of display. Their existence changes a UK domestic space into one where Africa as a continent is inscribed as part of the valued and elevated experiences of the owners of the home. The home is contextualized through the place of Africa, and the objects are contextualized through their presence in England. The value of the object in remembering Africa is simultaneously elevated reconstructed.

> **Darshna:** In your home do you have pieces that are always with you wherever you move?
>
> **Bharti:** Not religious, but some of them like the carvings of Kenya ... I still have them. Those were the pieces that we always have in the home. Masai figures, animals ... the furniture I have carved, I bring from Kenya.
>
> **Lalita:** In Mombassa, even if you go out, you know, on the beach or go to the streets you see people selling them, you know. On the streets, kiosks, full of these carvings, and I think that is a kind of trademark of Kenya. And I see them and it reminds you of Mombassa ... Mombassa? The memory is beautiful ... But then I come back home. ... That's why I had to get very Asian furniture.

There is a sense from some women that these *located* textures are essential, and that even if a variety of objects is collected to make a home, the 'Indian' aesthetic has to dominate. Encounters with animals are recounted as epic experiences by the women. In recalling the landscapes, the elephants and other animals are amplified in their size and colour and are super-dynamic in the way in which they are recalled. Their size, strength and speed are vivid and made vital through the objects. This is not surprising considering the iconography of Kenya and the Rift Valley in general; for residents, the African National Parks were immediately accessible. Set off against the ecology of the UK, these memories and experiences become heightened and magnified. They are imbued with sentiment, pleasure and pride until the objects that are refracting these moments are themselves elevated.

The objects and their own aesthetics are valued beyond their material worth and function. These objects acquire value through a form of devotion, of reverence of the memories they signify. Through time, these materials shift in meaning; ecologies become concentrated in iconic form, distorted, extended, and enhanced through new contexts and different moments.

The possession of goods, materials and objects often consolidates the enfranchisement to those landscapes of heritage that is refracted through the objects themselves. The owning of these objects and the *belonging to* enhance the memories that the object bears. The relationship with materials is sometimes figured through a sense of loss of land or a way of life; distortion occurs through displacement. The sentiment of *loss* signifies the temporal as well as spatial distance from these meaningful landscapes. Through memorialization, these figure act as mediators between times of lived experience and a sense of East African heritage that contributes to British Asianness. The textures of these curios are as significant as the images in them. Handcrafted Mvuli wood furniture, coasters of animals in the National Park, copper etchings of elephants and ivory bracelets make the sensory memories collide with the experience of being in England. They also evidence the necessity of connecting back to these landscapes, which secure a common social history. For many of the women, the connection made through material cultures is described as internal, as *biological*; a connection through gut, blood, and consciousness. Possessing these curios is a way of possessing, retaining access to, and enfranchisement to this past, both lived and remembered as a landscape of heritage. The act of appropriation gives back life to the object; in this new 'reconstruction of value' (Geary, 1986), it activates connections with this past heritage not only through being a relic of a lost landscape, but also as a regenerative embodied gateway to memories of other lives and societies.

Souvenirs of Ecological Memory

Whilst the group discussions were valuable in giving a social context and an insight into social meanings that resonate amongst the women in relation to these material and visual cultures, the home interviews ensured a grounded sense of the women's relationships with different places and environments. I was able to situate their relationships with visual culture at the point of their 'presencing' in the home. Some of these constitute symbolic landscapes of belonging, and others are more immediate memories of home, but all are *agentic* and not simply reflective. The women situate their identity (*triadic*) between geographical coordinates of the present, past and utopian landscapes of belonging. The essential quality that these cultures have is their metonymical value, as they not only trigger places framed in the women's memory, but they also trigger all the other smells, textures and sounds of that place and moment. In Susan Stewart's (1993) terms, we can position these cultures as souvenirs which refigure the temporality within which we live and abstract the culture of nature that the souvenir makes present. Nature, landscape, and the very geographies of the souvenir are abstracted, made sublime, and become a sacred gateway to a displaced sense of a past nature and culture, shifting *this* nature and culture through its presence.

> All souvenirs are souvenirs of nature, yet it is nature in its most acculturated sense which appears here. Nature is arranged diachronically through the souvenir; its synchrony and atemporality are manipulated into human time and order. (Stewart, 1993, p. 150)

These visual and material cultures operate as pervasive mnemonic and active devices in the process of identity reconstruction. They act as 'memory maps' and operate in constructing a formalized politics of national heritage and history (Slymovics, 1998); these cultures are therefore integral and not peripheral to British national culture. Memory-maps, like the diaspora, are mobile and transportable, and they allow for a continued relationship with a cultural past, but a dynamic lens of shifting territorial identity politics situated within new cultures of nation and citizenship. The cultural materials and visualized landscapes presented here, together, express the everyday experience and values of landscape in the women's sense of Englishness. They display the nature of the lens through which the textural modes of citizenship are experienced within the home and beyond. The mobility of the objects is central to their shifting meaning and value to the British Asian women. The objects themselves have their own social biographies, which get re-contexualized and in turn reorder their present in the context of the new site of home (MacCannell, 1992).

> Biographies of things can make salient what might otherwise remain obscure. For example in situations of culture contact, they can show what anthropologists have so often stressed: that what is significant about the adoption of alien objects – as of alien ideas – is not the fact that they are adopted, but the way that they are culturally redefined and put to use. (Kopytoff, 1986, p. 67)

Kopytoff (1986) focuses on the cultural signification and recontextualization of Western objects situated in Africa. Here, I am dealing with the reverse of this, materials that have come from India and East Africa to Britain. They have their significance centred on 'other' cultural landscapes, and have travelled to the centre of empire. As materials, they carry a lot of cultural capital; they signify the greater cultural body of meaning far beyond their value and use. They act as metonymic, artefactual devices (Csikszentmihalyi and Rochberg-Halton, 1981), acting as a reference to the whole country, culture or biography, but they are just a splice of that memory, experience or culture, and are lodged into other memories in the UK. Their reconnecting is done from the present, and these things hold meaning privileging the contemporary context. It is this metonymical value that I consider in relation to visual cultures in the home. It is their power to signify, refract and translate complex geographical relationships that is examined.

Visuality/Materiality: Placing the Visual

Visual cultures in the British Asian home, such as photographs, fabrics, pictures, and paintings, have meaning and value beyond their textual content. The objects of visual culture considered here *presence* the landscapes of South Asian migration, thus importing 'other' landscapes previously shaped by colonial governance (Drayton, 2000) into a British context. There is a movement and circulation of landscape imagery which reflect post-colonial experiences of living in colonial landscapes in East Africa and South Asia. There is a need within cultural geography to attend to the materiality of the visual cultures that we engage with beyond their text and aesthetic form. The texture of a visual object such as a photograph, print, or fabric dynamically informs the viewers' interpretation of the image. The scent, touch, and prismatic quality of the materials of culture were the primary texture of engagement with it, or were certainly equal in value to that of the text. Essentially, the material texture of the visual artefact had a critical significance. These visual cultures refract, represent, and are metonymical signifiers of other environments and landscapes. They also refract sensory engagements with other places, landscapes and natures. Shards of other environments are enclosed in these visual cultures. In the domestic space, a collage of other environments is produced through the display and collection of visual cultures in the home. They are significant in their material presence in that they ground identification in tangible and textural engagements. Their materiality of the visual is an extension of anthropological interest in the biography of material cultures, and the nature of domestic cultures in connecting across temporal and spatial axes of lived experience (Appadurai, 1986). If the materiality of the visual is an additional register of the text, then we need to extend research into the way that material cultures operate on the scale of the visual; the sighting of material textures is as valuable as their being situated within a spatial matrix (Holt and Barlow, 2000; Tolia-Kelly, 2001).

For migrant populations that have traversed several landscapes to take up residency in Britain, the experience of migration from a previous home has meant the forced discarding of objects, photographs, clothes, documents and furniture in the task of uprooting and re-rooting. For many of the women, the biographies of the material cultures were themselves just as significant, as they had traversed these landscapes with them. When they had obtained secure accommodation after migration, the value of those few objects preserved was enhanced in the process of their appropriation and display. Some of these items were purchased on trips back to Kenya, India or Pakistan, as souvenirs which in turn were reconstructed through a lens of distance and loss. Social anthropologists (Csikszentmihalyi and Rochberg-Halton, 1981; Appadurai, 1986) and cultural geographers (Rose, 2000, 2003b) have written on the value of material and visual cultures (including photographs) within the domestic sphere. All have contributed to the understanding of the way that these cultures are products of relations that extend beyond the home (Rose, 2003a, p. 5). The values and meanings assigned to these visual cultures indicate the women's connection with these refracted landscapes of belonging, which are critical in securing an enfranchised space of 'home'. The material nature and biography of these visual cultures are considered here alongside the shift in their meaning in their new sites of display after migration.

The Photo-Object

As a visual culture, photography operates in different ways in the social and cultural contexts of the home. As a medium, it engages with a different 'way of seeing' from other forms of visual media; gazing at photographs is very different from viewing a film or a painting. My central concern here is the value of the photograph beyond the textual imprint embedded in the photographic paper (Pinney, 1995; Ryan, 1997). In the context of South Asian post-colonial migration, the preservation of family photographs is limited; often these types of materials are left behind or shipped. Often they are peripheral to other essential items needed in the process of resettlement. The effect of having photographs in these circumstances enhances their social meaning and value. They record real moments and events, resonating with a memory of relationships in particular geographies. The photographs are fragments of real biographies. However, the photographs symbolize broader oral histories and personal relationships with people and places. They become symbolic of these places and the social histories in these places; Knappett (2002) describes photographs as simultaneously iconic and indexical: metaphor and image. Imagination and the text become conflated, imbuing the photograph with narratives beyond the image. Multisensory experiences are recalled through the text of a photo – the heat of the sun, the scent of the jasmine flower, the feel of the humid air, and other impressions. Photographs also trigger memory from the imagination of other 'texts' or textures; through them, family narratives and recollections of a past life are evoked. In the home, their effect is to reconnect with other landscapes and places of enfranchisement. Fragments of remembered landscapes are lodged in the image through symbols, aesthetics and textures. These photographs operate as metonymical devices which trigger memories of a nation, an intimate garden or a sense of self connected to oral narratives of the past. Over time, these photographs themselves attain 'relic' status; they are revered and treasured. This is exemplified in Sheetal's relationship with the photograph of a Tanzanian landscape mentioned below.

Sheetal brings with her a photograph that she describes as *dil no tuklo*, 'a piece of my heart'. The photograph is not a representation of this landscape, but a part of her core identity. Sheetal was brought up in Tanzania in the 1960s and 1970s. Her grandfather's family had moved there as a response to the incentives offered to Indians by the British colonial government in India to develop the economic landscape of the East African protectorate. Sheetal, like many Asians in East Africa, moved to Britain in 1972 and now lives in a large family home in Harrow. Sheetal's reverence of this photograph was so intense that I did not, as I had wanted, take the photograph away to copy it. It would have been unethical to risk losing the photograph in transit; its loss would have had the effect of mutilation. Her relationship with the photo demonstrates some of the real, imaginary and symbolic values with which the material of the photograph is imbued. The text itself is an image of the Bismarck Rock (later named Mwanza Rocks) on the lakeside of Lake Victoria. This image is intimately connected to her sense of self, which is dependent on this piece being in place, in her home.

> **Sheetal:** This is from my hometown in Mwanza. This has been our focal point. All my family and almost everybody from Mwanza has this place as their memoir. ... Yeah, loads of memories. And everybody had their special thing with Mwanza Rocks ... you know, something connects. Yeah, central to almost anybody if you ask from Mwanza, they will have something to talk about. Where the rocks were, it was intriguing every time we went there. The rocks are perched so delicately on each other. ... This was a place of leisure where the whole family would travel there to have a picnic and watch the sunset. ... This has been our focal point ... all my family and almost everybody from Mwanza has this place as a memoir.

The photograph is preserved in a plastic cover. It is displayed in the intimate space of her bedroom. The plastic cover indicates that it is not fixed *in situ*. There is an incongruity between the reverence Sheetal has for it and the flimsy covering of shiny plastic which obscures the scene. At the sight of the photograph, Sheetal recalls the social life of her family enjoying picnics at Lake Victoria. Along with visualizing this scene, she remembers the heat of the summer, the scents and sounds of food cooking, children playing and the rush of the water. Inscribed within her recollection are sensory textures which add dimension to her mind's eye image of this past landscape, but which also operate as an embodied memory; they are experienced as a sensory recollection. This memory is fixed in a sensory experience in a particular space-time (Edgerton, 1995; Sutton, 2000). The recall of a physical memory is that of an experience beyond the formal documentation of the past. The textures recalled are beyond the parameters of the photograph; they are operative as independent triggers of the memory of this scene. The refracted memories are given coordinates through these biographical landmarks, which assist the interpretation of events, thoughts, and sensations from the past, in the contemporary sphere. These are a material store of sensory experiences not part of a linear geographical route, but of a collage of sensory of identifications (Tolia-Kelly, 2001).

The photograph transports Sheetal back to the rocks; the text takes her to the memory of Bismarck Rocks, to a place where she is able to 'connect' and explore her own mental and emotional imaginary. This was a place that allowed her to contemplate and wonder; there is a sense of freedom in her narrative linked both to her adolescence and to her distance from Tanzania. This was the place where she had her first dates, her first kiss, and also the place where family relationships were strengthened through group activities. The activity, through memory, gets polished and embellished with positive events; the memory is airbrushed, smooth and shiny. Redolent of a rose-tinted adolescence, family life, and a place of connection and belonging, East Africa is symbolized by the image of the Bismarck Rock:

> **Sheetal:** This is like families would go down there. Couples would sit down there. There's my dad and mum with their friends, and there was numerous hours of fun and talk. *Shanti* ['peace']. Five o'clock you would go there and not finish until it went dark. There are three or four good hours. *Akho time nikali jai* ['All time would pass away'].

The photograph becomes a means of 'being' within this powerful place in Sheetal's memory. The experience of migration changes this from a vital memory of this experience to a memoir, a place that is active only as a memory of these feelings of connection and an embodied wholeness. The photograph becomes a store of social history rather than a place vital and present that can be reconnected with. These, in the context of Sheetal's description, resonate with her belief that there is no chance of returning to Tanzania to as a permanent resident. The memories of this social life are also engaged with the practices of making a home here and her social life. This image is displayed in Sheetal's homespace. Here the domestic landscape in Britain is contextualized through Tanzania; real moments of lived experience in Britain are figured through her nostalgia and this symbol of a past landscape.

Family Frames

Figure 5.6 shows a photo of Kajal as a young girl with her brother and his new wife. At the sight of the photograph, Kajal was in tears. She hasn't seen her brother or her nieces and nephews in years, as she has not been back to Bangladesh in years. The photograph refracts the memories of the landscape in Kajal's village, Balishastra (the nearest town is Molobibazaar), as it was before her migration. Kajal is a British citizen who is intimately connected with the landscape of Bangladesh; her migration to Britain situates her inbetween a belonging to a now imagined landscape of Bangladesh and the experience of living in Harlesden in Brent, London. Through distance, the photograph become symbolic of a pre-migratory landscape of home, which reminds her of her loneliness and marginality in Britain. Through time, this image has become a symbol of Kajal's intimate emotional relationship with this place. The picture itself is very formal; the backdrop is a classic iconographical landscape in a photographer's studio. There is a river or lake in the foreground and palms to the right. There are flowers and clouds annotating the classic scene. This type of luscious scenery as backdrop is not uncommon in Indian studio photography, as it conforms to a very particular set of rules (Pinney, 1997); the formality of the sitters is in tune with the formality of the backdrop. The colours in the photograph are intensified and the forms are defined heavily by the brightness and solidity of the colours worn. There is a hyperintensity that is

Figure 5.6   Kajal's family

imbued in the image through aesthetics. The photograph has been sent to her by her brother and sits in her living room, next to images of an Islamic site of *Hajj* (pilgrimage) at Medina. The proximity of the two indicates the reverence she has for this photograph. The image of Medina also indicates her alternative cultural identification with the landscapes of Islam.

Kajal describes Balishastra and her family, which are framed within the photograph, with emotion. The picture is a memorialization of her family, not just those within the frame, but also those excluded. Through it, Kajal talks through her family history and their current residences. The term for taking a photo in Gujarati and Hindi is to '*ketch*', meaning to take or to draw. When translated, it seems that there is a sense of taking something away from the scene and set in the material of the photograph. The image has drawn from the sitters, but also from the landscape of Bangladesh. Pinney (1995) describes how, within this belief system, the photograph is revered actively as an icon of the life of the person. As individual portraits are made (funerals, weddings, births), these portraits become the story of that person, and encompass the being of that person. On prayer after death, rituals and rites are performed on the photograph of the person in the same way as rites and rituals are performed on chromolithographs of religious icons. Pinney (1995, p. 111) argues that most Indians

> have a number of old images which continue to accrue potency as they become accreted with marks of repeated devotion – vermilion tilaks placed on the forehead of deities, the ash from incense sticks, smoke stains from burning camphor.

For Kajal, this image is a direct tracing of a significant and material moment. The physicality of photography is exactly that, it is a tracing of the light reflected on the solid materials in the frame. The family image is taken from that reflection in the lens. This photo is revered because of the tracings of that moment, but also because it has been elevated from that moment, made potent through the reverence that Kajal gives it. This reverence in the everyday imbues the photo with a value beyond a sentimental record. Pinney describes this process as giving the photo 'breath'; the photo evolves its own life through the context of display, but is imbued with a 'soul' through the reverential way in which it is treated by Kajal. These photos are singular, they are not numerous and therefore are more precious. For Kajal, this is a piece of her family 'taken' from them, and watched over by her. Her relationship with the image is about the recording of a past moment, but, more crucially, it is to keep this moment alive through making it potent in the practices of everyday life in Britain.

The photograph is a social record of the fads of that time. The clothes, carpeted floor, and lush, green, fertile surroundings also indicate wealth and prosperity. The backdrop exists to ensure a statement about position and success but also to record a moment in a fixed way. The formality is deliberate. The family members do not show emotion: they are there to record their history and social connections. This purpose has been played out in Kajal's relationship with the photo. Her own sadness is a reminder of her brother's distance from her, but also of her own marriage break-up. Her family has been broken and she has been failed by the promises of the perfect framing of families within this genre. The photograph is symbolic of family networks and moral living, at the same time as being symbolic of Bangladesh and the ecology of Bangladesh. Through gazing at the image, she is reminded of her very real blood relations, memories which are superimposed on imaginary narratives about nation, family and marriage. The image is symbolic and is a record of an event; it captures a way of life in Bangladesh. In the group discussions, Kajal describes the summers of fear when tigers threatened the villagers; she describes the nature of the surroundingthick forest and the vulnerability of the village to heavy rainfall, endangering the harvesting of crops. The picture holds a relational importance to living in Britain. It links Kajal to Bangladesh and her past citizenship there. This relational identity is constantly affirmed in Britain in the process of applying for state benefits and public housing as a single parent; in all of these official documents, she is asked to state her country of origin, and her first language, all of which link her to her Bangladeshi origins.

African Dioramas/English Idylls

A picture of zebra (Figure 5.7) is owned by Shanta. It is framed and placed centrally in the main family room. For Shanta, incorporated within this one image is a connection to her biographical route to Britain, her first home in Kenya, and her subsequent settlement in Malawi. While gazing at the image, Shanta describes the scents, sounds and tastes of picnicking near Lake Naivasha in Kenya, and seeing zebra on the journey to Lake Malawi; for her, the landscapes of these two nations conflate into a singular iconography of hyper-real animals, jungle, flamingos, and the dry, dusty savannah. Her family have several of these paintings by David Shepherd, a wildlife artist (painting and portraiture) whose work began in Kenya. He is well known for his paintings of the African savannah. Shepherd's images represent an iconography of African landscapes and nature, and he has contributed to a singular vision of African landscape being about wild animals and native people, resonant of a colonial lens. David Shepherd has been an active environmentalist and received the OBE for his services to wildlife conservation in Africa. He is also an honorary Fellow of the Royal Geographical Society. His images are hung in Shanta's living room, bedrooms and dinning room. They are placed there as fragments of East Africa reflecting into the home personal histories of East African life. The images present an African aesthetic and incorporate the narratives of East African life into the day-to-day living space through their aesthetics and imagery. There is a collaging of 'English' and 'African' aesthetics, lace, modern fittings, pastel walls, overlaid with zebra, giraffes and elephants from the savannah. These materials together reflect the varied connections that Shanta and her family have with the landscape of East Africa as encountered in the London borough of Harrow. The seeming incongruity reflects the multiplicity of cultural connections that reflect Shanta's migratory experiences; these resonate in her home, reflecting her culture of being a post-colonial migrant settled within the leafy avenues of suburban Pinner, Harrow.

Figure 5.7   Painting of zebra in Shanta's home

Boy Fishing is a mass-produced poster in Shilpa's home. This was pointed out as 'special' by Shilpa in a tour of her home. It is hung in the entrance hall to her flat. It has pride of place where every visitor can see it. It's a surprising, sugar-sweet, greeting-card image of a rural idyll, a childhood idyll, in fact, certainly aesthetically situated in a European or English landscape. At the sight of this image, Shilpa is reminded of her uncle's home in Uganda, along the river. She describes the freshness of the scents, the rippling of the river, and the pleasure of being free to roam in this landscape. She recalls the games they used to play and how safe she felt in the area. For Shilpa, this image is made meaningful in her flat in Harlesden. It is positioned in Britain as a testament to the landscape of dreamy childhood days in Uganda – of a luxurious home, a luscious countryscape and the pleasure of free roaming. The aesthetics of the image are chocolate-box pastel, a saccharin ode to a fantasy of a sentimental, picturesque scene, incongruously located in Uganda in Shilpa's narration. This image highlights the meanings that material cultures have beyond their text as well as the nature of the circulation of landscape meanings in colonial history. The image of an English pastoral in this story is cross-cultural; it is translated as an aspiration to a picturesque scene, which embodies pleasure, peace and a sense of innocent childhood pastimes. However, the image is a familiar one and in the tradition of landscape representation; it is not ideologically neutral (Bermingham, 1994). It embodies a vision of English landscape that is seemingly benign and nostalgic; the image reiterates a culture based on a fantasy of nature and childhood. What it occludes is the social politics of the icon of a white, playful child in England, a representation that is an impossibility for Shilpa's own child now living in England. *Boy Fishing* also evidences the power of the English pastoral in the South Asian imaginary, a legacy of living within a British colonial state and culture prior to migration.

The scene, like many others displayed in homes across Britain, is valued partially because it embodies a scene contrary to that of urban living, especially in the inner London Borough of Brent where Shilpa is situated. It also becomes a respite from the struggles of resettling and bringing up a daughter single-handedly after migration. Shilpa contrasts her life at her uncle's to her difficulty of finding employment in the UK. She has a BSc in chemistry from Mumbai University, but for the last two decades has been unemployed. In this context, the image holds more than a scene of Ugandan pastoral; it also harbours a landscape of hope beyond the struggle of isolation and alienation experienced in her day-to-day life. Shilpa's visits to the women's centre are her sole means of access to a social network. Her engagement with *Boy Fishing* also shows how refractions of 'other' landscapes are not always triggered through an exotic African palette. As viewers, we imbue images with socially and culturally specific visual vocabulary; we make meaning through our codes of reading, signifying and interpretation. The contexts of display are also critical in activating these textual meanings. *Boy Fishing* denotes an English pastoral that is folded into a series of narratives about the nature of England and a visual iconography of Englishness. The image initially seems incongruous and unexpected. After I talked through its value and resonance with Shilpa, it was clear to me that the text of the image was meaningful beyond the registers and visual vocabulary of the genre. The material of the image can be read through Shilpa's biography. The water in the image becomes equatorial: the grass becomes savannah. The scents and sounds shift from a European textural space to a Ugandan one. In the context of Harlesden, the matter of the image operates as a gateway into a past landscape of Uganda, but also into a social record of past landscape textures relevant in their absence or non-attainability. Lived landscapes, utopian landscapes and England are juxtaposed on the surface of her flat wall. The entrance hall in the Harlesden flat is sometimes synchronized with the landscape in Uganda, but is made meaningful in relation to being and living in Britain. Shilpa's narrative and the embodied memories are present in the matter of the poster; the poster becomes a third space where Uganda, and England merge through a process of refraction, and reflection.

Sheetal's *Bismarck Rock* displayed in her home in Harrow also starkly contrasts with the suburbanscape of public parks and estates of semi-detached houses. The intensity of attachment is triggered through the tension between being here in Britain and the intensity of attachment to a life in Tanzania, which cannot be reclaimed, as living in Mwanza is not possible. The text's ability to transmit those textures into her home now offers a form of suture. A piece of her very being is on display, ensuring that Northwood in Harrow, London, is given meaning through the presence of Lake Victoria, not just the image but the scents, tastes, touch of African life. These refractive textures from visual cultures on display combine to consolidate connections to other landscapes. Their new contexts of display give these cultures a new 'cultural vitality' (Gell, 1986, p. 86). They inject scents, sounds and textures of other landscapes into the British home. Woven together, these aesthetics represent a territory of culture, a territory collated together to support a sense of belonging and being that makes sense of migratory journeying and telescopes these textures to create a place of settlement and roots. To some degree, these cultures contribute to the memorialization of 'other' people and places; they become artefacts of a biographical journey as well as a social history of the group. These visual markers are tokens, souvenirs of another country, another landscape, which at the time of their lived experience was shaped by British colonial rule.

Conclusions

As I have argued in this chapter, lived environments harbour the precipitates of *re-memory* as they figure as narratives of social heritage. These solid precipitates, in the form of visual and material cultures, help situate diasporic groups politically and socially within 'structures of feeling' that have evolved through their varied relationships with national identity. These identifications have infinite configurations but continue to have a connective cultural significance amongst the South Asian diaspora in the UK. *Re-memory* adds to these geographical coordinates by incorporating social history, not as directly experienced, but as forming part of the cultural identity narratives which are live within the diaspora. A true understanding of the post-national, post-colonial experience is enabled through an interrogation of 'home' landscapes and their material cultures. These are a critical contribution to understanding social geographies of migrant communities, which shape and reshape the social, cultural and political landscape in Britain. The South Asian diaspora's sense of boundedness is disrupted through processes of uncovering discontinuities in these differences in the roots and routes of the community, as well as the experience of an unsustainable biological and cultural integrity. East African Asians are neither African nor Indian. Therefore, for them, any formulation of a 'national' or 'traditional' national culture is dependent on the domestic scale, the creative space where diverse configurations of 'Asianness' can be found. These material cultures, at home, represent the physical buffer between their experience of displacement and the difficulty in dealing with marginalization from new points of settlement. These cultures operate as a psychic investment in a set of 'textures of identification'; they reflect these transnational communities' shared 'structures of feeling'. These cultures situated in the homespace offer a sense of inclusion, which has aesthetic, sensual, and psycho-sociological dimensions. They are active in their ability to locate contemporary British Asian identity in the context of post-colonial geographies of migration; English*ness* is in a continual process of being remade and reformulated. This is not a new mode for the culture of Englishness, which is and has, in turn, been situated throughout the colonial territories and rememorialized through experiences in the physical landscapes of

England. The cultures of landscape through which we recognize *Englishness* have shifted historically; the geographies of Englishness thus are neither unitary nor exhaustive. These are deeply geopolitical geographies of Englishness negotiated in the everyday lives of migrants in Britain.

Bibliography

Appadurai, A. (ed.) (1986), *The Social Life of Things: Commodities in Cultural Perspectives* (Cambridge: Cambridge University Press).

Appadurai, A. and Beckenridge, C. (1992), 'Museums are Good to Think: Heritage on View in India', in I. Karp, C.M. Kreamer and S.D. Lavine (eds), *Museums and Communications: The Politics of Public Culture* (Washington, DC: Smithsonian Institution).

Araeen, R. (1992a), 'Cultural Identity: Whose Problem?', *Third Text* 18, 3–5.

Araeen, R. (1992b), 'How I Discovered My Oriental Soul in the Wilderness of the West', *Third Text* 18, 86–102.

Araeen, R. (1991), 'The Other Immigrant: The Experiences and Achievements of Afro-Asian Artists in the Metropolis', *Third Text* 15 (Summer), 17–28.

Bermingham, A. (1994), 'System, Order, and Abstraction: The Politics of English Landscape Drawing Around 1795', in W.J.T. Mitchell (ed.), *Landscape and Power* (Chicago: University of Chicago Press).

Bloch, M. (1954), *The Historian's Craft*, trans. R. Putnam (Manchester: Manchester University Press).

Boym, S. (1998), 'On Diasporic Intimacy: Ilya Kabakov's Installations and Immigrant Homes', *Critical Inquiry* 24 (2), 498–524.

Brah, A. (1999), 'The Scent of Memory: Strangers, Our Own and Others', *Feminist Review* 61 (Spring), 4–26.

Connerton, P. (1989), *How Societies Remember* (Cambridge: University of Cambridge Press).

Csikszentmihalyi, M. and Rochberg-Halton, E. (1981), *The Meaning of Things: Domestic Symbols and the Self* (London: Cambridge University Press).

Drayton, R. (2000), *Nature's Government: Science, Imperial Britain and the 'Improvement' of the World* (New Haven, CT: Yale University Press).

DuBois, W.E.B. (1994 [1903]), *The Souls of Black Folk* (New York: Dover Publications).

Edgerton, S.H. (1995), 'Re-membering the Mother Tongue(s): Toni Morrison, Julie Dash and the Language of Pedagogy', *Cultural Studies* 9: 2, 338–63.

Fanon, F. (1967), *Black Skin, White Masks* (New York: Grove Press).

Fanon, F. (1961), *The Wretched of the Earth* (Harmondsworth: Penguin).

Fanon, F. (1959), *Studies in a Dying Colonialism* (Harmondsworth: Penguin).

Ferguson, R., Gever, M., Minh-ha, T.T. and West, C. (eds) (1990), *Out There: Marginalisation and Contemporary Cultures* (Cambridge, MA: MIT Press).

Finney, B. (1998), 'Temporal Defamiliarization in Toni Morrison's "Beloved"', in B.H. Solomon (ed.), *Critical Essays on Toni Morrison's 'Beloved'* (London: Prentice-Hall International).

Gandhi, L. (1998), *Postcolonial Theory: A Critical Introduction* (St Leonards, NSW: Allen and Unwin).

Geary, P. (1986), 'Sacred Commodities: The Circulation of Medieval Relics', in A. Appadurai (ed.), *The Social Life of Things: Commodities in Cultural Perspectives* (Cambridge: Cambridge University Press).

Gell, A. (1986), 'Newcomers to the World of Goods: Consumption Among the Muria Gonds', in A. Appadurai (ed.), *The Social Life of Things: Commodities in Cultural Perspectives* (Cambridge: Cambridge University Press).

Gilroy, P. (1993a), *The Black Atlantic* (Cambridge, MA: Harvard University Press).

Gilroy, P. (1993b), *Small Acts: Thoughts on the Politics of Black Cultures* (London: Serpent's Tail).

Halbwachs, M. (1992), *On Collective Memory*, ed. and trans. by L.A. Coser (Chicago: University of Chicago Press).

Hall, S. (1990), 'Cultural Identity and Diaspora', in J. Rutherford (ed.), *Identity: Community, Culture, Difference* (London: Lawrence and Wishart).

Hirsch, E. and O'Hanlon, M. (eds) (1995), *The Anthropology of Landscape* (Oxford: Oxford University Press).

Holt, Y. and Barlow, P. (2000), 'Visual Culture in Britain: An Introduction and a Debate', *Visual Culture in Britain* 1: 1, 1–13.

hooks, b. (1995), *Art on My Mind: Visual Politics* (New York: New Press).

hooks, b. (1994), *Outlaw Culture: Resisting Representations* (London: Routledge).

Horovitz, D. (1998), 'Nameless Ghosts: Possession and Dispossession in "Beloved"', in B.H. Solomon (ed.), *Critical Essays on Toni Morrison's 'Beloved'* (London: Prentice-Hall International).

Karp, I., Kreamer, C.M. and Lavine, S.D. (eds) (1992), *Museums and Communications: The Politics of Public Culture* (Washington, DC: Smithsonian Institution).

Knappett, C. (2002), 'Photographs, Skeuomorphs and Marionettes: Some Thoughts on Mind, Agency and Object', *Journal of Material Culture* 7: 1, 97–117.

Kopytoff, I. (1986), 'The Cultural Biography of Things: Commoditization as Process', in A. Appadurai (ed.), *The Social Life of Things* (Cambridge: Cambridge University Press), pp. 64–94.

Lambert, S. (2001), *Irish Women in Lancashire 1922–1960* (Lancaster: Lancashire Centre for North-West Studies).

MacCannell, D. (1992), *Empty Meeting Grounds: The Tourist Papers* (London: Routledge).

Mehta, R. and Belk, R.W. (1991), 'Artifacts, Identity, and Transition: Favorite Possessions of Indians and Indian Immigrants to the United States', *Journal of Consumer Research* 17 (March), 398–411.

Mercer, K. (1990a), 'Black Art and the Burden of Representation', *Third Text* 10 (Spring), 61–78.

Mercer, K. (1990b), 'Welcome to the Jungle: Identity and Diversity in Postmodern Politics', in J. Rutherford (ed.), *Identity: Community, Culture, Difference* (London: Lawrence and Wishart).

Miller, D. (2001), *Home Possessions: Material Culture Behind Closed Doors* (Oxford: Berg).

Mitchell, W.J.T. (ed.) (1994), *Landscape and Power* (Chicago: University of Chicago Press).

Modood, T. (1994), 'Political Blackness and British Asians', *Sociology* 28: 4, 859–76.

Modood, T. (1990), 'Muslims, Race and Equality in Britain: Some Post-Rushdie Reflections', *Third Text* 11, 127–34.

Morrison, T. (1990), 'The Site of Memory', in R. Ferguson, M. Gever, T.T. Minh-ha and C. West (eds), *Out There: Marginalisation and Contemporary Cultures* (Cambridge, MA: MIT Press).

Morrison, T. (1987), *Beloved* (London: Chatto and Windus).

Parkin, D. (1999), 'Mementoes as Transitional Objects in Human Displacement', *Journal of Material Cultural* 43, 303–20.

Pinney, C. (1997), *Camera Indica: The Social Life of Indian Photographs* (London: Reaktion).

Pinney, C. (1995), 'Moral Tophilia: The Significations of Landscape in Indian Oleographs', in E. Hirsch and M. O'Hanlon (eds), *The Anthropology of Landscape* (Oxford: Oxford University Press).

Rose, G. (2003a), 'Just How, Exactly, Is Geography Visual?', *Antipode* 35, 212–21.

Rose, G. (2003b), 'Domestic Spacings and Family Photography: A Case Study', *Transactions of the Institute of British Geographers* 28, 5–18.

Rose, G. (2000), 'Practising Photography: An Archive, a Study, Some Photographs and a Researcher', *Journal of Historical Geography* 26: 4, 555–71.

Rutherford, J. (ed.) (1990), *Identity: Community, Culture, Difference* (London: Lawrence and Wishart).

Ryan, J.R. (1997), *Picturing Empire: Photography and the Visualisation of the British Empire* (London: Reaktion).

Sale, M. (1998), 'Call and Response as Critical Method: African-American Oral History Traditions in "Beloved"', in B.H. Solomon (ed.), *Critical Essays on Toni Morrison's 'Beloved'* (London: Prentice-Hall International).

Samuel, R. (1994), *Theatres of Memory* (London: Verso).

Slymovics, S. (1998), *The Object of Memory: Arab and Jew Narrate the Palestinian Village* (Philadelphia: University of Pennsylvania Press).

Solomon, B.H. (ed.) (1998), *Critical Essays on Toni Morrison's 'Beloved'* (London: Prentice-Hall International).

Stewart, S. (1993), *On Longing: Narratives of the Miniature, the Gigantic, the Souvenir, the Collection* (Durham, NC and London: Duke University Press).

Sutton, D.E. (ed.) (2000), *Remembrance of Repasts: Materializing Culture* (Oxford: Berg).

Tolia-Kelly, D.P. (2002), 'Iconographies of Diaspora: Refracted Landscapes and Textures of Memory of South Asian Women in London', PhD thesis, Department of Geography, University College London.

Tolia-Kelly, D.P. (2001), 'Iconographies of Identity: Visual Cultures of the Everyday in the South Asian Diaspora', *Visual Culture in Britain* 2, 49–67.

Williams, R. (1979), *The Fight for Manod* (London: Chatto and Windus).

Wilson, A. (1978), *Finding a Voice: Asian Women in Britain* (London: Virago).

Chapter 6

Mediations in Memory/History: The Art of Making Environmental Memory Tangible on Canvas

Tangible Memory

In the case of writing on memory, the nature of archival, textural and narrated texts structures the argument and the possibilities of research completely. For post-colonial migrants, such as South Asian women, the materials are scarce and thus the methodology used in the process of research is innovative and designed for this group's particular geographical and historical positioning. Ogborn (1999, p. 104) makes it clear that 'accounts of the making of memory and heritage inevitably become the reconstruction of geographies.' In this research, geographies of Britishness are being remade, negotiated, and made sense of through memories of pre-migratory landscapes that are valued as formal iconographies of past citizenries. The plurality of memory in temporal, spatial and cultural registers and forms is also negated in the works of these memory theorists. The binary between individual memory narratives and collective ones is reasserted in their writings. The critical work on nostalgia within geography (Blunt, 2003; Legg, 2005) has enabled me to rethink the productivities of memory, in particular nostalgia in a post-colonial realm, thus enriching notions of memory from oral history or Proustian accounts of sensory memory. Other writers such as Huyssen (2003) recognize the political problems of asserting the binarism of history as being 'authentic' and memory purely as 'selective' and fallible. He is also concerned with memory beyond the testimonial and the traumatic and argues for attention to be paid to the transformative and changing registers of memories in our urban space. Here, I propose that memory can be made tangible; in a visual form, the canvas becomes the site of memory/history. Memory/history transcends the testimonial; it is inclusive of mobile memories, exhibited in public space, thus challenging both the usual form of memory/history as (written) text, and the usual representations of geographical history. Landscape painting and the form of textual representation are *part* of the story here: an edited version of the women's voices, memories and imagined geographies of race, landscape and citizenship.

For Post-Colonial Landscape Memories

For some readers, there is a critique to be made of the 'relevance' and the need to use 'landscape' in research with everyday people. In my research so far, my project has been to acknowledge the international circulation and dynamism of the figure of landscape across and between cultures, especially those which have been under British colonial rule (see also Jazeel, 2005). For artists such as Zarina Bhimji, thinking 'nation' is intrinsically linked to colonial visions of landscape and the resulting post-colonized territory. Imagining the nature of the new colonial territory involved thinking about these spaces through the double-faced project of emulating Britain abroad through 'improvement' (Drayton, 2000) as well as thinking of them as 'Oriental' (Said, 1978), 'tropical' (see Driver, 2004, on Brazil), and ripe for British 'ordering' (see Blunt, 2003, on India in the 1930s). Landscape is a feature of the colonies – their social, economic and cultural geographies. Being and living in the colonies are about thinking identity through a lens of the imperial landscape vision. The colony is designed, aesthetically and materially, to fit the blueprint of governance which reflects landscape values of the empire, but also the active enfranchisement of the imaginations of the colonized; they imagine themselves within a culture of landscape citizenry embedded in lived socio-spatial territories of belonging. An example of how links with landscape are encountered and engaged with despite exile from East Africa is given by Zarina Bhimji (2003) in *Out of Blue*, a film of post-colonial landscape in Uganda, shown at the Tate Britain in 2003.

> The opening scenes of Zarina Bhimji's *Out of Blue* reveal the breathtaking landscape of Uganda. However, almost immediately this luscious vista is disturbed by the murmur of voices and the crackle of flames. The film shows various places which suggest elimination, extermination and erasure. Many Asian and African residents were expelled from Uganda by General Idi Amin on 9 August 1972, events which provide a background for *Out of Blue*. They are also part of the history of this country, since many of them came to Britain in the early 1970s to start a new life.
>
> *Out of Blue* can be seen within the tradition of British landscape painting, as it captures the mood and historical significance of a place through representations of the countryside. The intense and atmospheric soundtrack includes the natural sounds of birds, fire, and echoes from the buildings filmed. (Tate Britain website)

The colonized were politically enfranchised, as citizens, through their labour in 'improving' the national landscape. Therefore, it is misguided to position the post-colonial, imaginative realms as being *outside* the landscape tradition. This is part of the political discourse of this research; 'landscape' is in circulation, in the cultural imaginary of the mobile British, including those colonial officers who set out with their visions of Britishness within the empire. Any notion that 'landscape' should be retained as an imaginative figure of cultural meaning for the colonizer or Western body is a symptom of what Gilroy (1992, p. 57) terms 'cultural insiderism':

> An absolute commitment to cultural insiderism is as bad as an absolute commitment to biological insiderism. ... We must beware of the use of ethnicity to wrap a spurious cloak of legitimacy around the speaker who invokes it. [Culture] is fluid; it is actively and continually made and re-made.

Many British Asians have experienced double and triple migration. Recording these remembered landscapes is a way of locating South Asian history and 'positioning' in Britain. This is a particularly marginalized community often excluded from academic research (Bhopal, 2001). This research seeks to elucidate their transnational relationships; however, it does not assume that this new transnational positioning is implicitly empowering or emancipatory (Pratt and Yeoh, 2003). The women's landscape images and their oral histories reveal a sense of loss, disempowerment and nostalgia for past environments that is both productive and negative. Oral history itself has challenged conventional history in its assumptions, values and biases, but it also has been stretched by expanding the very notion of what counts as legitimate archives or artefacts. There are alternatives to the archives and artefacts normally presented in museums and collections in formal spaces of displaying heritage and history. Frisch et al. (1990), in their review, have shown that museums of history and sites of heritage themselves have materially become the points of tension and contestation that have led to new histories emerging. This evidences the extent to which communities themselves are actively engaged in creating and representing their own history. Biographical narratives can therefore be enriched beyond narrative by thinking beyond linear texts, and 'storytelling'. Here, the aim is to capture by a multimedia method the aesthetics and the sensory mode of environmental memories. For historical geographers of the marginalized, it is important to be aware of the political and ethical implications of collecting, collating and disseminating these histories. It is not enough to talk through and re-present the transcripts; there is a need for ethical reflexivity and racial sensitivity (see Kyriacou, 1993).

The 'Describe a Landscape' collaboration (Figure 6.1) offers a set of methodological and conceptual engagements with memory to try to expand contemporary geographical and memory research.

Methodological Practice and Visual Mediations

The research process employed is an intervention using geographical visual methodology as a means through which South Asian women's transnational, landscape values can be recorded and pictured in the public sphere. Geography

Describe a landscape . . .

You are invited to a selected show of paintings by Melanie Carvalho. The paintings are the result of a collaboration between the artist, the women of the Harlesden Asian Women's Resource Centre, the women's group from the Harrow Sangat Centre and Divya Tolia-Kelly, who is a researcher in Cultural Geography at University College London

The exhibition runs from Tuesday 2 January – Friday 5 January 2001
in the Sherwell Gallery,
University of Plymouth - Admission Free

Melanie Carvalho and Divya Tolia-Kelly will be speaking at the RGS-IBG Conference on Thursday 4 January in Module 3 of the Social and Cultural Geography Research Group Session (14.30-16.00) on 'Visual Cultures and Geographies of Identity'

This show is sponsored by Ecumene and is part of a larger project conducted by the artist with financial support from the London Arts Board

ECUMENE

Figure 6.1   'Describe a Landscape' project poster

is often positioned as a visual discipline, from the creation of visual media to the way that geographical epistemologies are embedded within a vocabulary of visual metaphors (Sui, 2000) and to the centrality of visual tools to communicate geographical knowledges (see the debate between Mike Crang, Gillian Rose, Felix Driver, David Matless and James Ryan in *Antipode*, volume 35, 2003). In this debate, Rose (2003) quite rightly asks us to reflect on the creation of theatres of the geographical within the spaces of the dissemination of research; she calls for the reconsideration of the hierarchies that emerge through the ocular mechanisms of advancing geographical thought. The use of the visual is political, and thus the employment of visual media and, in turn, visual methodologies has to be undertaken with some reflexivity (and modesty).

The thinking through of landscape as a visual register of identification is not a new intervention, but positioning the visual as a means through which to grapple with memory in processes of identification is. In this stage of the research process, we aimed to disturb the usual hierarchies of viewing, and the 'scopic regimes' that structure both the gallery space, and the painting of landscape on the canvas itself (Rose, 2001). This investigation is premised on the literatures on landscape as providing 'ways of seeing' ourselves within discourses of nation (Daniels, 1993; Matless, 1998), heritage (Lowenthal and Prince, 1964, 1965; Lowenthal, 1979, 1985, 1991), and 'the parish' (Crouch and Matless, 1996). Landscapes of memory/ history are made tangible in the 'Describe a Landscape' project through the form of its canvases and transcripts. Post-colonial Englishness as much as a transnational landscape citizenry is reflected in these forms. The sessions with the Asian women were designed and conducted with the landscape artist Melanie Carvalho. This was in contrast to the traditional orientation of landscape research, and was thus informed by writers such as Berger (1972), Kinsman (1995), Mitchell (2002), Nash (1994, 1996, 1997) and Rose (1997). Exhibiting these women's engagements with landscape onto canvas in the formal spaces of the gallery also made the gallery space inclusive of landscape values of post-colonial migrants.

The methodological process was very productive. Initially, there was an issue of language and meaning, which manifested itself most in the use of the term 'landscape'. Carvalho has been trained in the Western art tradition at the Central Saint Martin's College of Art and the Royal College of Art, and as a scholar at the British School at Rome; thus, the canvases were informed by these intellectual conventions. These conventions were sometimes pitted against the women's own landscape imaginary.

> **Shazia:** But landscape doesn't just mean sand, sea, mountains, hills. It can mean streets even …

> **Carvalho:** Yes, street scenes, er, I suppose these aren't strictly landscapes. They are sort of, this is only a couple, there are a whole series of beaches. … Now living in a city as well it's quite difficult to sort of paint landscape, because it's something unknown to you and you don't really, you're not really familiar with it.

This exchange illustrates how Carvalho excludes the urban when defining landscape. In the women's own theorizations, landscape is not situated in the rural. For other women, there is a more formal, nationalistic definition of the word 'landscape'; the word 'matabhumi' was used in the session by some women; this word is a direct translation of *motherland*, and has nationalistic connotations.

In the 'Describe a Landscape' sessions, Carvalho gave a slide talk of her work and explained her biography and interests in landscape painting. She showed the women slides of descriptions and paintings already done as part of the 'Describe a Landscape' project. The women were invited to participate, and it was made clear that this was optional. During the exercise, the women sat in a group whilst drawing; Carvalho and I joined them in this process. The women annotated their pictures with writing. Before the women started their sketches, I asked them to close their eyes, relax, and then, in their mind's eye, to travel to a place that they imagined as their ideal place of home. In this technique the women were led to reconnect with their landscapes of enfranchisement. Images of places, of friends, and of childhood pleasures came back in rich profusion. The value of this is to uncover valued environmental memories. I did not emphasize 'childhood' or 'environment' as themes, but did use a technique to relax the women to a state of letting their minds focus on their visualization.

The memory/history that resonates in the descriptions, in their aesthetics, forms and landscape scenes, reflects multiple layers of meditation, ending with the canvases themselves. In this set of descriptions and canvases are imprints of actual gardens, plants, buildings and landscapes, contextualized through the annotations. Whilst writing and drawing, they talked about homes in Lahore, Gujarat, Bangladesh and Kenya, and their sketches symbolized these places. Memories made up of particular objects, plants, buildings, and climates were put together in a collage, combined to form a singular image from multiple memories of the colour, shape and texture of these environments. The places from the women's imaginations came alive on the page; this was a gateway into these other landscapes, including aesthetics, colours and textures, and creating another imagined impression. These descriptions are a set of tangible environmental memories that the women have that constitute a way of being elsewhere, geographically, culturally and historically. Through the process of creating the descriptions, the women create a historicized visual image, but one which resonates with all their senses. These are also multi-temporal, not fixed on a singular moment. From each of the descriptions, Carvalho produced a representation from her own interpretation, in oil, on canvas, as a means of materializing the imagined landscapes of home further. Carvalho treated these as technical 'instructions' to paint. She wanted to be as true to the descriptions as possible, as she describes:

> The use of painting as a medium is important in the work because it is also about
> how landscape painting as a genre plays a part in mythologizing ideas of where
> we're from and who we are. From the descriptions, I made landscape paintings
> on a landscape-format canvas in oils, all traditional materials,... The painting
> has its own presence and it doesn't belong to me and it doesn't belong to the

person that made the description. I try to do the paintings in one sitting. The idea is to be as objective as possible and respond technically as possible, without elaboration. To create a landscape painting from someone else's imagination is really exciting. (in Anderson et al., 2000, p. 115)

The Women's 'Describe a Landscape' Descriptions

The interpretation of images (see Appendices 1–15) made by the women and Carvalho was done through a coding process in which I looked for the relationship between imagined landscape and the actual biography of the women. The coding of these led me to analyse these visual descriptions in relation to their memorializing of the local ecologies of depicted places. I looked at the 'sites' and 'modalities' as summarized in publications on visual methodologies (Leeuwen and Jewitt, 2001; Rose, 2001). In the analysis of the descriptions made for Carvalho, I focused on expressing their 'compositional interpretation', which became a means of coding them. These codes were contextualized in relation to their production (in a group interview process) and the biography of the producer.

Overall, the women's descriptions are tracings of mental images, and their compositions reflect environmental memories in different ways. There are place-specific landscapes recalling a landscape which is real and remembered. These are like testimonies of places which collage actual aspects of places in the women's memory. I have also shown a set of descriptions in which the women give an impression of an ideal landscape, which evokes sensory memories, recorded on the page as impressions of the aesthetics and textures of that place. These are not attempting to trace or record the actual landscape characteristics but give a sense of place through the description. Childhood memories are co-present in all of these images.

Carvalho's painting is made from Chandan's sketch (Figures 6.2 and 6.3). In this, Chandan locates her landscape in the memory of a moment, and thus the description is a place-specific collage. This has iconographies placed within the frame that together make a real remembered place. This is not time-specific, but is a historicized imaging of memories of a singular place over time. In this description Chandan sketches a set of water structures including a well, a water pump, a water tank and a swimming pool within the same frame. These are figured around a house, which has a garden path, and a tree, which has palm-like leaves but is labelled 'almond tree'. The sun is high in the sky. Time is figured in the image as a sequence denoting the progressive development of access to and domestic use of water. The use of water was originally limited to a well. This well has a small *loto* (copper carafe) attached to it, indicating limited access but also economy of use. The water pump is next to it. Both are in the front of the scene. They are more immediate memories. As we move back, the water tank is attached to the house. It is evidence of wealth and the overcoming of limited access and availability of water. To the right of the frame, we see the pool – an indulgent, superfluous pool. It is coloured

Describe below, in writing, or a drawing, or however you like, the landscape that best represents your idea of home.

Please return to: Flat 7, 51 Tudor Road, London E9 7SN

my Ideal Home IN DHRANGADHRA
in District of Surendranagar Gujarat State
in India in 1968 – 1976

Figure 6.2   Chandan's description

turquoise blue. It is like a vision. The blue is significant; it is vibrant and significant in its presence. The colour and the size indicate its dream-like luxury. Chandan states, 'This is my ideal home in Dhrangadhra in the district of Surendranagar, Gujarat State, India, 1968–1976.' It is the pool, which is imagined, and this is a futuristic projection, setting up a tension between past scarcities and future excess. Indulgent and unnecessary, the existence of a pool in the Kutch desert would be an 'ideal' because, in reality, the heat and extent of the dry season would make it a fantastical project. This is an area where water is revered because of its scarcity. That the pool sits alongside the well indicates the temporal projections of the cultural practices and values of water. The pool can only be imagined from the position of an outsider, and the projection is made possible by migration. Water, historically, has not been scarce in England and so the possibilities of a pool in the Kutch desert

Figure 6.3   Carvalho's painting based on Chandan's description

are imaginable. The sketch, therefore, is framing a way of seeing this domestic landscape in the context of the development of infrastructure, access and socio-cultural practices. Water, from being a necessary resource, becomes a material of play. It is an aesthetic and material presence which lifts the description from a time-space specific memory. The landscape is a collage of water iconographies, which are a historical sequence of the materials that Chandan has used and memorialized in relation to India. They are emblems of place, culture and a valuesystem for water, which shifts in relation to migration. At the exhibition held at University College London, Chandan offered her response to Carvalho's painting:

> **Chandan:** First of all there was a well so we had to take the water from the well to put water in the home. Then the municipal they put the tap water. A tap. A public tap! And then they put a tank near my house.

> **Carvalho:** So that's quite funny. We've recorded the different stages at once! So how does that feel? Does it seem like the place you made or does it seem like a different place to how you imagined it?

> **Chandan:** It's fine. Yeah [nodding vigorously].

The painting of Shilpa's description (see Figure 6.5) is an example of how a place-specific description evokes an ecological memory. Shilpa's description (Figure 6.4) evokes a scene in Sudan; its dryness and lack of water are evident. The heat of the sun and lack of vegetation, ponds and plants except for date palms are illustrative of the desert climate. This Sudan scene is different from a childhood scene; it is recorded as a specific view from a window in her home in Sudan. Shilpa describes a way of life. This is recorded in the frame. The focus is on the difference in the shape of the houses, which are built as round mud huts, with dried grass on the walls and roofs. The image is like a tourist photo; it has nothing of a personal nature within it. The image is an iconographic overview of a scene, rather than a detailed, intimate childhood landscape. This memorialization flattens the image, and the items within the frame are designed to indicate the order of the landscape

Figure 6.4   Shilpa's description

as a visual record. This description is a testimony to Shilpa's environment in Sudan, and the world she looked out on from within the confines of her home. As a woman, she did not enter the settlement in daylight unless her husband accompanied her. Women only left their homes in the period before sunrise to collect water for the family. In terms of memory/history, Shilpa's painting records the presence of South Asians in Sudan in this colonial period. Also it reflects her memory of this ecology as discordant to the landscape of England, as encountered in the inner-London borough of Brent.

In these paintings, the remembered places of the past are remembered through different ways of seeing the past. Sometimes in the images it is clear that these are scenes of a childhood home or a nostalgic view of a specific experience. The past can also be read by the iconographies within the frame. Water and the way it is depicted illustrate this. On the beach, or at a swimming pool or at a lake, these scenes are leisure scenes in which the women idealize the pleasure of these water features in their past. Water is also present in a domestic context; a water pump, tap or tank provides a reflection on the development of the country at the moment the image shows. In the idealized scenes, water is available freely, and exists in an exoticized form.

Figure 6.5   Carvalho's painting based on Shilpa's description

Impressionistic Environmental Memories

Kokila gives an impressionistic description (Figure 6.6) that reflects the essence of the place, or the sentiment of connection that the woman has to the place. In this image, there is a faint line on the page. The feintness allows for an impression to develop in the viewer's imagination. The scene is suggested: the tree and the river are non-specific. There are no indications of location, scale, colour, or grounding. But, in her description, Kokila evokes a sense of place, which is an impression of peace and sublimity. This is a scene of prayer at the Subramati river in Ahmedabad. The only well-defined object is the prayer pitcher, which is the *loto* (copper carafe). This object is positioned here as symbolic, in contrast to Chandan's positioning of it as functional. Here there is an absence of detail, but this absence inscribes a sense of emotional and mental clarity: the moment of devotional prayer and sacrifice. The tree is not described as any particular type, but it is drawn like a tree in England – an ash or a birch. This is indicative of the

Describe below, in writing, or a drawing, or however you like, the landscape that best represents your idea of home.

Please return to: Flat 7, 51 Tudor Road, London E9 7SN

Figure 6.6   Kokila's description

Figure 6.7   Carvalho's painting based on Kokila's description

position from which the tree is drawn; it is from elsewhere. The aesthetics is scenic and imbued with a sense of memory and reverence. Carvalho's painting (Figure 6.7) actually taps into the scenery without intrusively reworking it too much. The objects in the description are not collaged or layered, but tightly joined in the memory of prayer at this riverside. The tree is symbolic of shelter, of nature, and fixes a location without grounding the scene with man-made structures. The river is idyllic because it is not fixed in shape or topography. It is simply suggested in a symbolic image of flow; an unpolluted, organic, life-giving force which is ritualized through Hindu religious practice.

> **Kokila:** When I pray, I remember this place ... remember how we must be still in prayer. Our people always used to pray like this ... our mind must be free.

Many childhood memories incorporated in the descriptions reflect physical sensory memories. The description drawn by Kanta (Figure 6.8) is an example of where the classifications are not simple. This impressionistic image was drawn by Kanta; in this description, Kanta fills the page with colour. There are only three identifiable things within the frame; the name 'Mombassa', a coconut palm, and a line that reads 'carefree days before coming to U.K. Aged 14'. The palm is iconographical in that it is used as a symbolic representation of a tropical idyll. Mombassa is the scene. The colours used are part of the idealization of African beaches. The aesthetics is more important than the iconographies, however. Reds,

Describe below, in writing, or a drawing, or however you like, the landscape that best represents your idea of home.

Please return to: Flat 7, 51 Tudor Road, London E9 7SN

Figure 6.8   Kanta's description

pink, purple, green and turquoise are juxtaposed to depict the collage of colours found in any paradise idyll. The evocation is of a place beyond objects but imbued with a sense of pleasure, freedom and sensuality. The line which annotates the image indicates the position from which Kanta remembers; it is from the place of being in England, with all the care and worries that the move has meant. These are assigned not only to being in the UK but also possibly to the burden of responsibilities that increase with age. The writing is part of the picture and the text is part of the aesthetics of the description; placing the text within the frame is reflective of the importance of written English in Kanta's experiences of living in England. Not only is there evidence of the negotiation of place and alienation from new locations, but also the core means of survival and expression is at stake. In England, Kanta was forced to eat meat and became very self-conscious about speaking English with an accent. The children in British East Africa were all taught English at school, a legacy of imperialism, but speaking this learned language in the UK was a source of humiliation and difference. For many migrants, in-betweenness also involves linguistic relationships. The impressionism of the drawing is an expressive tool. It creates placelessness with signification for certain

Figure 6.9   Carvalho's Painting based on Kanta's description

emotions and moments. Fixing and defining are secondary in these descriptions to the intellectual engagement with identity and non-locatedness of these emotions. The palm tree appears, but no other aspect of ecology is featured. This again is a placeless icon figured to locate a memory of another biome, not defined as a local place but as an experience of textures. These figured as a placeless experience allow for the description to become a dialogue about positionality, placelessness, and exclusion. Kanta's words, 'Understood without tension Tree flowing having to explain!!!! Acceptance', underline this positioning. There is a desire for acceptance mentioned in each of Kanta's descriptions. Home, for Kanta, is the space where she can be understood, be accepted and feel under no pressure. Her drawings are not traumatized expressions of alienation, but celebrate a memorialized past where a certain scene or set of colours can signify settlement and inclusion. These colours and iconographies are important in that they are familiar and intimate evocations of a sense of place which offers an inclusionary home where she has belonged.

Nostalgia and memory are embedded in all the descriptions, whether they are written or sketched. The effect of memory is to connect certain events and experiences from the past to the present. Nostalgia has the effect of tinting, smoothing and idealizing the past. Nostalgia adds a romantic glaze and elevates the memory from the grounded reality of the moment. Memory also serves to stretch and layer the past as it is informed by sensuous recollection and emotional meanings, and fed by oral histories and other media. The memory and its accuracy are contingent. The only way to trace the nostalgia with which the women fabricate their mind'seye view of the past is through their celebratory attitude to things in the country they lived in before migration. Childhood homes, in all the images, are decorated with positive, idyllic emblems; cultivated gardens, spaces for play ponds, lakes, swings and solid, two-storey houses.

Shanta's painting (see Appendix 11) is an example based on a written description. She describes a place called Thika Falls, where there are 14 waterfalls. This written description is annotated with a drawing of a gramophone. In the description, Shanta says that this is a childhood recollection from 9–11 years old. I discovered that Shanta's experiences of picnics at waterfalls and the lakefront were a constant in her life in East Africa. As a child in Kenya, she had gone to Thika, but later as an adult, she had visited Lake Malawi. In her home in Pinner, Shanta has photos of these picnics at Thika falls on display. The photos of these moments give context to her written description. They represent leisurespaces in East Africa, which Shanta visited in a convoy of family cars, just like in her pre-teen years. The effect of looking back is to record the place and a way of living in those years in East Africa. The description also frames cultural values and norms as lived out in these times. In many of the descriptions within both groups, it is evident that the idealized landscapes are imbued with versions of moral living. In Shanta's image, it is clear that this is a description of the whole extended family going to Thika falls in Kenya. These are her adolescent years. She describes this as 'childhood'. A scene of pleasure and fun is created. There is music, dance, singing, and eating. She describes travelling in a truck, the journey being part of this ideal landscape, and the textures also are important. The stones are cool, and smooth, the music invoking a joyous atmosphere. The scene is also a moral scene of family values expressed in a joined journey, with the picnic being as important as the journey. There is a wholesomeness reverberating throughout her description. People are swimming fully clothed. This is a subtle illustration of communal living, where there is a strict moral code. The house and garden ideal, which is rarely achieved in the UK, is embraced by many of the women as possessing moral value. Comments were often made about the way that living away from South Asia has influenced the moral conduct of their children. England is considered as having a modernizing but sometimes destructive influence on the acceptable practices of social behaviour and individual life choices. In the representations, relationships that are non-sexual are all evident, and partners are absent. These created landscapes are asexual and familial.

Childhood is the key thread in Zubeida's description too. She has a house and two children running towards the house. Their path is decorated with birds, flowers and grass. There is a large flower by the house, and the house is drawn like a face. Under the description Zubeida has written:

> Pakistan Lahore. My childhood was in this house, with huge garden, my dream is for my children to play in the huge garden. Children playing with swimming pool. Gulab, Chumeli – white flowers. Ducks – Jasmine smell. Weather – neither very hot or very cold. Orange fruit – tangerines.

The description in itself is a seemingly naively drawn, happy scene. The annotation describes where the drawing fits within Zubeida's imagination of an ideal home. There is interplay between a European sense of an ideal garden and an Asian one. Flowers such as jasmine (chameli), honeysuckle and roses all exist alongside ducks and tangerines. These seem discordant within the description; they are European iconographies placed in a South Asian context. The swimming pool, as also seen in Chandan's description, is an icon of luxury; its excess is evidence of wealth and an abundance of water. Whilst drawing this sketch, Zubeida talks about how she would like this scene for her children, a house in a sunny climate, with these flowers from her own childhood and an idyllic house. She projects her own childhood into the aspirations that she has for her own children. In her life in the UK, this past garden is still a source of her hopes for their future success. It forms a narrative about social mobility, an assertion of a set of values that were developed in the social context of Pakistan.

Many of the descriptions cannot be represented here and therefore I have decided to focus on a feature of two (see Appendices 4 and 8) which are formed in a particular way. I have grouped them together because of the splitting of the page that occurs in both descriptions. Puja (see Appendices 4 and 5) draws three lines dividing the page. Each section is part of a story of landscapes of home. The splitting of the page in these two sketched descriptions can also be analysed in terms of a splitting of the self, or of a multifaceted identification with place. In Puja's description, the

Figure 6.10   Carvalho's painting based on Zubeida's description

page looks like a set of scenes in a guidebook or a series of postcards. The top layer is a drawing of a house and kitchen-garden; there is a label 'mother's house'. The trees include papaya and *mitho limdo* (sweet neem), sometimes flowering. The second scene is of a cow and a calf and a thatched shelter for them. The third is a figurative scene showing Puja and her three brothers and sisters. The first layer signifies the importance of her mother's home in Puja's life. The domestic garden space is shown as an ideal family kitchen garden. This is a very practically drawn garden in that it includes a water tap, food, and the house structure and is recorded as a particular childhood home. The cows represent a sentimental memory, representing an attachment to a place in an idealized agricultural setting, as a cow in a shelter suckles a young calf. The splitting is a tool by which Puja represents three layers of her memory biographically; her family home, her siblings and the rural idyll which she treasures. There is no temporal splitting, but an identification split on the page between these relationships. In the second section, her siblings are the background for herself which she has drawn larger than the others. Here she is separate from them, and her distance from them is signified. Puja has not seen her siblings since the 1970s. She is identified differently in each scene. Her landscape of home is split to ensure a variety of connections and sentiments. For others, the splitting of the page ensures that multiple landscapes are represented, a collage of East African territories and British scenes. The sense of a cultural landscape of mobility, separation and social connection emerges. The artist's own process of mediation becomes visibly present in the canvases; it is embedded at every stage of production.

The Value of Carvalho's Paintings: What Do They 'Do'?

Embedded in this research process is a fixing of memory/history that is narrated by these Asian women.

It is inevitable that Carvalho's own values of landscape have influenced the composition, aesthetics and a particular landscape imaginary represented in the canvases. The way that Carvalho imagines landscape is that the composing of a landscape enables representations of temporal and spatial fluidity. The memories are embedded, imagined coordinates from which to compile current landscape positioning culturally and socially. The women's landscapes of home are inscribed with a place that is a collage of different sensory idealized ecologies of lush green plants of East Africa sitting alongside Indian and English ones. The pleasures of these different sensory moments are re-created by Carvalho to produce a collection which is a collage of these landscapes of citizenship. There is a multiplicity that is continuous in the women's descriptions. Some of the women themselves talk about landscape in the sense of a European definition understood commonly as an objective overview of a landscape scene which has limited presence of industry, the urban, or people. At the same time, there is a presence of specific places in their descriptions. One of the women comments on this:

Lalita: But I think that Mel's experiences had come in on the oil paintings as well. Particularly some of them, because when someone describes, say, a beach with coconut palms I think she can visualize quite easily the beaches that she's seen in Goa. Two of these pictures – I thought, this is Mel's picture of her place, and I didn't realize it was someone else's.

Visitor: Yeah, you couldn't do it any other way.

Carvalho: But that's obvious that you're going to have to put in your own ... because they're the only images you have of a place, then, your place. ... Often I did have to completely makeup. ... I mean they described trees that I didn't know, so all I had was their drawings. So I used their drawings of a tree but it's made up.

Carvalho's family background is Goan. Carvalho's identification has been a subject of tension between her and her tutors at art college. This experience whilst training also made her question this occlusion of complex identifications in favour of neat stereotypes. The 'Describe a Landscape' project, or her interpretation of the 'landscape' genre itself, ultimately allows for complexity, fantasy and multiplicity to be expressed in a formal, ideologically defined genre of art. During her time at Saint Martin's College of Art, Carvalho was encouraged to paint in a certain aesthetic because of her own identity positioning:

Carvalho: They love exotic imagery and they find it interesting. I always felt that it wasn't really me, and I did it just to please them. ... I thought English people don't make paintings about being English so why should I make paintings about being Indian?

In my view, Carvalho's interpretations onto canvas enlarge and multiply the sensory moments that are inscribed on the canvas. By mediating the descriptions onto canvas, new landscape scenes emerge. She inscribes a personal aesthetic into these landscapes of home. Carvalho acknowledges this process as a natural one that cannot be avoided, but tries to limit the effect of this by her aim to produce the painting in one sitting.

Carvalho: But that's so they're all painted wet on wet. That's the style of the painting ... it's much more impressionistic, or immediate.

The palette used is quite a distinctive set of colours and juxtaposition of colours. A distinct tropicalness is represented. Pinks and reds sit next to each other with greens and yellows. Carvalho paints this aesthetics through the descriptions, bringing them to life. The paintings are vibrant and colourful. They are also made to be contemporary, because of the glossy texture. This gloss adds newness to the images. It creates a sense of lightness that reflects not a faded memory or imagination but a vital one. This resonates with the women's descriptions of India, Pakistan and East Africa. The palette used by Carvalho resonates with the women where their environmental or biographical experiences are 'truly' pictured. This is about placeexposure, as opposed to a racialized aesthetics. This is a subtle relationship between colour and culture, in which cultural products and artefacts express a lived experience or memory of relationship with the aesthetics of the environment – of light, air, scent and sound. An aesthetic 'tropicality' resonates with Carvalho's canvases 'as a twin to the temperate' (Driver and Yeoh, 2000, p. 1). Carvalho and Cosgrove's discussion below reveals much about the circulation of the slippery and plural (Driver, 2004) notion of the tropical and the currency of its projection onto 'other' landscapes, known and unknown. There is an incoherence in the imaginative geographies of 'the tropics', which sometimes reinvigorate the homogeneity of the 'Oriental' world in aesthetic as much as cultural and natural terms. These relationships are being translated and expressed in the descriptions and are further transformed through Carvalho's own occularity. Professor Denis Cosgrove discussed this with Carvalho at the exhibition.

Exchange Recorded at an Exhibition at University College London, July 2000

Denis Cosgrove: When you get pink and red together against green and yellow it gives the impression of tropicality. ... you didn't have to use pink and red ... it's just a very Southern combination. When you see pictures of Cuba, there is always pink and red.

Carvalho: In India, people do have bright colours; they wear bright colours a lot. People did describe colours of flowers and I've just put those colours on in certain combinations which have created a tropical sense. Because they're like working drawings. They are pattern based.

In this project, environmental memories have been inscribed in the formal medium of fine-art painting. These memories 'presence' the experience of post-colonial migration in the sphere of visual culture, and its effects on identity and belonging in the UK for South Asians. The 'presencing' of these memories in this project has assisted the interrogation of the formal genre of landscape. The research project ensured that these memories were and are given a platform within formal networks of art exhibitions and art history. The women's memories and projected landscape ideals are materialized and recorded in the canvases. Carvalho's project is subversive in its political rupture of parochial and nationalistic landscape ideology; instead, landscape becomes an inclusionary and expansive medium. Identity is centred in the environmental memories of past landscapes. The centring of self in an idealized landscape gives the power of perspective, overview and ownership to the subject. It is no longer a genre of objectification and disempowerment of subjects; the post-colonial gaze of the women is presenced. The women's perspectives incorporate lost landscapes, nostalgic pasts, and their biographical histories. These images are inscribed with the strength of their diversity. Complex, multiple connections to landscape are possible within Carvalho's frame; it is beyond the reductive identity politics of conservative nationalism, identifying a truly lived cosmopolitanism.

After the exhibition in London, we gave all the women prints of the canvases. On subsequent visits to their homes, I saw that these held pride of place in living rooms and bedrooms. The landscapes that have been produced are revered. Memories are about a visceral, productive, post-colonial citizenship (see Blunt, 2003) in England. They are about making a 'home' despite obstacles and struggles; the scent of jasmine, the shade of the tamarind tree, the aesthetics of nature in the South, and the feel of the sun high in the sky are encountered with reverence. In Blunt's (2003) terms, these imprints are records of the ways in which landscape memory and nostalgia are crucial to a sense of thinking future diasporic living. The production of the images links directly to the production of lived space that is formed through a negotiation with nostalgia through materials in the home and the mind's eye. New landscapes of living are dialectically shaped and evolved through the nostalgia of the post-colonial British citizen. The remembered landscapes of belonging, fixed in the form of a visual image placed in the women's homes, give the process of enquiry a strange circularity. These prints become part of the visual cultures that are inscribed in the women's homes. These are images of memories, memories of memories, and conflations of continents and sensory experiences, all contained as part of a dynamic archive of diasporic journeys situated within their homes in England. They are also archival in their presence as they resonate with individual, social and cultural memories, operative as a memorial to South Asian diasporic history.

Private View, University College London, July 2000

> **Arti:** Yeah, to me you've captured, you've actually captured the emotional feelings that I had ...

> **Carvalho:** Well, that's quite apparent in your drawing though. All I did with yours was transfer the drawing to painting. And what about the transference from it into a painting? Does that change anything, it being painted?

> **Arti:** Yeah, it's more fluid for me, the painting, whereas this is more gritty, what I did.

> ----------------------------

> **Lalita:** I just love it. I just like it.

Carvalho: A lot of people were very specific but yours were a sort of imaginary of two places. ... So I gave it a sort of Northern light, because so many people had a tropical light. ... So that you couldn't really tell whether it was India or ...

Lalita: A landscape can be from anywhere anyway. A combination of several. My image is all about the lake. Because I made such a big link, and the hills! Beautiful. Lovely.

Figure 6.11   Private view with Melanie Carvalho at University College London

Figure 6.12   Private view with Melanie Carvalho at University College London

The research process highlights the tensions between memory and history. Just as oral narratives are partial and fallible record of events, the canvases show that recollections of environments are fluid and dynamic – sometimes iconographical, sometimes impressionistic. These descriptions and canvases record fragments of individual biographies and fragments of social history embedded within idyllic recollections. However, as a set of canvases, these paintings are valuable in their reflection of environments and the aesthetic textures of environments that are meaningful and relevant to the South Asian community. The paintings operate as biographical markers which signify the variety of life paths taken by the women, through various lands. These markers emphasize the heterogeneity of British Asians. Through the display of the canvases in the gallery, the landscape genre has been appropriated through the lens of the post-colonial imaginary. These canvases have been produced through a series of mediations through a series of lenses; the artist, the describer and the geographer. Through these stages, these refractions accrue interpretation, but also serve to reframe the individual processes of history/memory locked within their aesthetics and form. Their value as geographical and/or academic modes of knowledge and dissemination are contestable. As Pink (2001, p. 597) suggests, any 'developments in the practice of visual methods in reflexively comprehending research and representation and in seeking ways to allow the subjectivities of informants a space in academic texts that both empowers them and acknowledges the "fiction" of any ethnographic representation, indicate that visual methods are being "properly" developed.' In this research, I have sought to reflect on the multiple points of mediation and to create a space for the ethnographic nature of the production process of the paintings. What has been created here is a space for methodological reflexivity on the need for recording what is normally intangible (here in the form of landscapes of memory/history). What is not being claimed is that painting acts as a restorative medium which produces an historical record that is formed through registers of known, proven, and identifiable events and places in given moments. The memory/history displayed in these canvases demonstrates the relevance of landscape as a 'way of seeing' from the position of the post-colonial migrant.

Bibliography

Anderson, P., Carvalho, M. and Tolia-Kelly, D.P. (2000), 'Intimate Distance: Fantasy Islands and English Lakes', *Ecumene* 8: 1, 112–19.

Berger, J. (1972), *Ways of Seeing* (Harmondsworth: Penguin).

Bhimji, Z. (2003), *Out of Blue.* Tate Britain, co-produced by Documenta 11, 2002, http://www.tate.org.uk/britain/exhibitions/artnow/bhimji/default.shtm (accessed November 2007).

Bhopal, K. (2001), *Gender, 'Race' and Patriarchy: A Study of South Asian Women* (London: Ashgate).

Blunt, A. (2003), 'Collective Memory and Productive Nostalgia: Anglo-Indian Homemaking at McCluskieganj', *Environment and Planning D: Society and Space* 21, 717–38.

Crang, M. (2003), 'The Hair in the Gate: Visuality and Geographical Knowledge', *Antipode* 35: 2, 238–43.

Crouch, D. and Matless, D. (1996), 'Refiguring Geography: Parish Maps of Common Ground', *Transactions of the Institute of British Geographers* 21, 236–55.

Daniels, S. (1993), *Fields of Vision: Landscape and National Identity in England and the United States* (Cambridge: Polity Press and Cambridge University Press).

Donald, J. and Rattansi, A. (eds) (1992), *'Race', Culture and Difference* (London: Open University Press/Sage).

Drayton, R. (2000), *Nature's Government: Science, Imperial Britain and the 'Improvement' of the World* (New Haven, CT and London: Yale University Press).

Driver, F. (2004), 'Imagining the Tropics: Views and Visions of the Tropical World', *Singapore Journal of Tropical Geography* 25: 1, 1–17.

Driver, F. (2003), 'On Geography as a Visual Discipline', *Antipode* 35: 2, 227–31.

Driver, F. and Yeoh, B.S.A. (2000), 'Constructing the Tropics: Introduction', *Singapore Journal of Tropical Geography* 2: 1, 1–5.

Frisch, M. (1990), 'The Memory of History', in M. Frisch (ed.), *A Shared Authority: Essays on the Craft and Meaning of Oral and Public History* (Albany, NY: State University of New York Press), pp. 8–23.

Gilroy, P. (1992), 'The End of Antiracism', in J. Donald and A. Rattansi (eds), *'Race', Culture and Difference* (London: Open University Press/Sage).

Huyssen, A. (2003), *Present Pasts: Urban Palimpsests and the Politics of Memory* (Stanford, CA: Stanford University Press).

Jazeel, T. (2005), 'Nature, Nationhood and the Poetics of Meaning in Rahuna (Yala) National Park, Sri Lanka', *Cultural Geographies* 12: 2, 199–227.

Kinsman, P. (1995), 'Landscape, Race and National Identity: The Photography of Ingrid Pollard', *Area* 27, 300–10.

Kyriacou, S. (1993), 'May Your Children Speak Well of Your Mother Tongue: Oral History and the Ethnic Communities', *Oral History* (Spring), 75– 84.

Leeuwen, T.V. and Jewitt, C. (eds) (2001), *Handbook of Visual Analysis* (London: Sage).

Legg, S. (2005), 'Contesting and Surviving Memory: Space, Nation, and Nostalgia in *Les lieux de mémoire*', *Environment and Planning D: Society and Space* 23, 481–504.

Lowenthal, D. (1991), 'British National Identity and the English Landscape', *Rural History* 2, 205–30.

Lowenthal, D. (1985), *The Past Is a Foreign Country* (Cambridge: Cambridge University Press).

Lowenthal, D. (1979), 'Age and Artefact', in D. Meinig (ed.), *The Interpretation of Ordinary Landscapes* (New York: Oxford University Press).

Lowenthal, D. and Prince, H. (1965), 'The English Landscape', *Geographical Review* 55, 186–222.

Lowenthal, D. and Prince, H. (1964), 'The English Landscape', *Geographical Review* 54, 309–46.

Matless, D. (2003), 'Gestures Around the Visual', *Antipode* 35: 2, 222–26.

Matless, D. (1998), *Landscape and Englishness* (London: Reaktion).

Meinig, D. (ed.) (1979), *The Interpretation of Ordinary Landscapes* (New York: Oxford University Press).

Mitchell, D. (2002), 'Cultural Landscapes: The Dialectical Landscape – Recent Landscape Research in Human Geography', *Progress in Human Geography* 26, 381–89.

Nash, C. (1997), 'Irish Geographies: Six Contemporary Artists, Nottingham', Exhibition at the *Djanogly* Art Gallery, University of Nottingham.

Nash, C. (1996), 'Reclaiming Vision: Looking at Landscape and the Body', *Gender, Place, and Culture* 32, 149–69.

Nash, C. (1994), 'Remapping the Body/Land: New Cartographies of Identity, Gender, and Landscape in Ireland', in G. Rose and A. Blunt (eds), *Writing Women and Space: Colonial and Postcolonial Geographies* (New York: The Guilford Press).

Ogborn, M. (1999), 'The Relations Between Geography and History: Work in Historical Geography in 1997', *Progress in Human Geography* 23, 97–108.

Pink, S. (2001), 'More Visualising, More Methodologies: on Video, Reflexivity and Qualitative Research', *Sociological Review* 49, 586–99.

Pratt, G. and Yeoh, B.S.A. (2003), 'Transnational (Counter) Topographies', *Gender, Place and Culture* 10, 159–66.

Rose, G. (2003), 'Just How, Exactly, Is Geography "Visual"?', *Antipode* 35: 2, 212–21.

Rose, G. (2001), *Visual Methodologies* (London: Sage).

Rose, G. (1997), 'Situating Knowledges: Positionality, Reflexivities and Other Tactics', *Progress in Human Geography* 21: 3, 305–20.

Rose, G. and Blunt, A. (eds) (1994), *Writing Women and Space: Colonial and Postcolonial Geographies* (New York: The Guilford Press).

Ryan, J.R. (2003), 'Who's Afraid of Visual Culture?', *Antipode* 35: 2, 232–37.

Said, E.W. (1978), *Orientalism* (Harmondsworth: Penguin).

Sui, D.Z. (2000), 'Visuality, Aurality, and Shifting Metaphors of Geographical Thought in the Late Twentieth Century', *Annals of the Association of American Geographers* 90, 322–43.

Tate Britain [online] http://www.tate.org.uk/britain/exhibitions/artnow/bhimji/default.shtm (accessed 16 March 2010).

Thompson, A., Frisch, M. and Hamilton, P. (1994), 'The Memory and History Debates: Some International Perspectives', *Oral History* 25th Anniversary Issue, 33–43.

Chapter 7
Post-Colonial Ecologies of Citizenship

At the heart of this research monograph is an ecological approach. Spatiality, power and politics are embedded in this research model, in which there is a critical intersection between power, citizenship and national identity. Without knowing the spatiality of post-colonial experience, we cannot think geopolitically about citizenship or indeed cultural identity. Geography has been implicit in theories of environmental determinism and scientific taxonomies of race, which in turn informed colonization and cultural determinism. If we are to study race in contexts of power, materialities and history, geographies, landscapes and nature are at their heart. Ecologies of living signify modern connections and senses of belonging to a nation.

These are liquid, modern times, always mobile, always transcultural and cosmopolitan. Mobility and the dynamism of landscape connections are woven through the biographies, material cultures and visual cultures produced here. In Massey's (2006) terms, landscape is in motion, in a geological time frame. Landscape is dynamically shaping present cultures of citizenship and national identity. The mobile cultures of the post-colonial landscape have been expressed here as a story of *Landscape, Race and Memory* which materializes *ecologies of citizenship*. Stasis in notions of race, ethnicity, culture, landscape, identity or nature is a myth, and thus a cultural approach to citizenship and cultural geographies of identity should be based on accepting and understanding our society's mobile cultures of citizenship. A need to embrace liquidity in art theory, praxis, landscape, culture and emotion can stand as a testament for future cultural worlds. This openness to cultural expressiveness and the nature of mobile cultures would and could result in a truly transcultural approach to landscape and art. In turn we can develop further archaeologies, histories, museologies, and academic scholarship which are truly ecological, embedded in an inclusive, ethical philosophy of research.

In this book, I have considered the role and value of visual and material cultures in figuring British Asian identifications with landscape. Landscape is positioned here as a material signifier of identification with land, territory and environments that contribute to formal and informal connectedness with national cultures and citizenship. British Asians' citizenship is figured through the experience of their residence in colonial territories within South Asia (including India, Pakistan, Bangladesh and Sri Lanka) and East Africa (including Kenya, Uganda and Malawi). South Asians' experience of these lived landscapes and their particular political status within colonial territories influence their connection with landscapes in Britain and the formation of 'Britishness'. Landscapes are represented, refracted and memorialized in the form of visual cultures within the British Asian home.

Visual cultures of landscape are situated within the South Asian as critical modes of securing a sense of being and belonging within Britain for this group of post-colonial migrants. Moreover, visual cultures, in the British Asian home, such as photographs, fabrics, pictures, and paintings, have meaning and value beyond their textual content. This research is an exercise in reading visual and material cultures through a materialist lens, which allows for an examination of their place in the embodied practices of making a 'home' in Britain. The objects of visual culture considered here *presence* the landscapes of South Asian migration, thus importing 'other' landscapes, previously shaped by colonial governance, into a British context. There is a movement and circulation of landscape imagery, which reflects post-colonial experiences of living in colonial landscapes. The presence of these landscapes in visual and material cultures shapes South Asian domestic spaces and illustrates the value of landscape to post-colonial residents of Britain.

Overall, there is an increasing need within social research to attend to the materiality of the visual cultures that we engage with. This research demonstrates the value of investigating this material dimension, through the process of researching domestic landscapes of the post-colonial migrant. These visual cultures refract, represent, and are metonymical signifiers of other environments and landscapes. They also refract sensory engagements with other places, landscapes and natures. Shards of other environments are enclosed in these visual cultures. In the domestic space, a collage of other environments is produced through the display and collection of visual cultures in the home. They are significant in their material presence in that they ground identification in tangible and textural engagements. Their materiality of the visual is an extension of anthropological interests in the biography of material cultures, and the nature of domestic cultures in connecting across temporal and spatial axes of lived experience. If the materiality of the visual is an additional register of the text, then we need to extend research on the way that material cultures operate on the scale of the visual; the sighting of material textures is as valuable as their being situated within a spatial matrix. We also need to consider the visual beyond the occularcentric, to include the embodied, material, and affective register of encounter and interpretation. The aesthetics of the material cultures in the home forms part of their sensory vocabulary, which in turn needs some attention.

Material and visual cultures are positioned here as active shapers of post-colonial identification with landscape. Their active place in the imagination of geographical relationships is examined through examples. However, their power is not limited to the domestic sphere. For almost all of the women in the study, the landscapes they had engaged with prior to migration have sustained relevance in their current lives. The presence of these visual landscapes translates African, Indian and other Asian experiences of landscapes into creating a set of familiar textures in this new domestic scene. Britishness and British landscapes are changing as a result of the migration of peoples and their landscape imaginaries.

Post-Colonial Britishness/Englishness

This book has delved into the corners of English life, although they are very different from those encountered in a book entitled *Landscape and Englishness* (Matless, 1998). Primarily, this is because questions of mobility and race have been incorporated into the theoretical orientation of *Landscape, Race and Memory*, using a post-colonial lens. However, Englishness remains as a constant presence within any post-colonial study of landscape, but is constructed through a notion of self and Other:

> This book is happy to delve into obscure corners of English life, but it proceeds from the assumption that a definition of Englishness as insular or unitary would not only be undesirable but also impossible to sustain. National identity is regarded as a relative concept always constituted through definitions of Self and Other and always subject to internal differentiations. (Matless, 1998, p. 17)

Landscape, Race and Memory has delved into these imaginative geographies of post-colonial Englishness, in practice. Englishness in motion, or Englishness as figured through 'mobility', is at the heart of making tangible British Asian memory-histories of their cultural experience. For British Asians, cultural relationships with landscape are dialectically linked to the route of migration and a sense of continued *diaspora*, whilst being located within British landscapes. Mobility is central to the visions of enfranchisement and belonging, as these sites of settlement are always in the process of being made in relation to moving away from other sites and locales. The role of visualized landscapes and material cultures as presented here is that through *their* mobility they enable a cultural vocabulary that communicates a mobile cultural citizenship. Yet, through their material presence they provide stability in the moment of being *English* and diasporic. These two modes are not mutually exclusive because an attunement to mobile cultural nationalisms is made possible through their presence and dynamism. However, I do not wish to promote the notion that all diasporas seek a bounded cultural nationalism located within a single territory of citizenship, but I argue that the desire for a sense of belonging and inclusion posits culture, nation, and/or a connection with a territory of culture as being a foundational mode of negotiating day-to-day living for all citizens. A diasporic cultural nationalism, then, is a source of stability within discourses of cultural heritage structured through the experience of mobility, migration and a type of transnationalism not bounded through the usual socio-cultural structures of nationalism. At some moments, this is imagined as a bounded 'other' citizenship located in 'Africa' or 'India'; at other moments, it is hybridized into new vernacular cultures of being English formulated through the diasporic imaginary.

The process of creating landscapes of enfranchisement through remembering lived landscapes is part of the process of positioning within a territory of culture that is secure and inclusive. Within the process of collecting this research, there was a sense that the Asian group had a set of common territories not bounded within a narrative of nation state or a singular territory, but within a *territory of culture*. The mapping of the landscapes in the previous chapters is not collectively signifying a single nation or citizenship, but announces a relationship with a set of landscapes that are resonating within the group as spaces of enfranchisement. In combination, they show how mobility and landscape cohere to form a landscape of belonging, through imagination, memory and experience. In everyday life, these material cultures are also valued as a shard of past connections to territory and nationalism; this is after mobility has occurred. Often the meanings of things are sedimented aftermobility, and in response to alienation and racism, and/or in the mission to 'make-home' and belong.

The colonized Indians who were later East African British secured British residency and thus homes after migration. They were dwelling in a new environment, ecology and biome very different from the ecology of the African Rift Valley and the savannah landscape. However, in East Africa these landscapes were not 'owned' in a sense of citizenship – they were experienced by subjects of British rule and the British protectorate. Their orientation towards this landscape was experienced as a member of an imported cultural group, set aside as different from Kenyans, Ugandans and Zimbabweans. Through mobility, these landscapes were appropriated, literally through material ownership of splices of the landscape – wooden carvings, zebra skins and touristic curios. As a mobile group, British Asians are marginal to national discourses, resulting in moments of disenfranchisement. In this situation the diaspora seeks to secure a sense of identity, belonging and new geographies of enfranchisement. The renewed, yet refigured attachment to colonies is made meaningful through the sense of exclusion, marginality and disenfranchisement experienced in England. These are the British-English cultural landscapes. These landscapes are studded with shifting sovereignty, the promise of 'improvement', and a dynamic and relational definition linked to England and empire. The post-colonial British landscape is international, mobile and a potent source of respite and nourishment in the context of the experience of race and racism. The experience of mobility is a catalyst for new mobile nationalisms. As mobile nationalisms, they are not embedded within firm nation states, but within a set of cultural modes of identification. These networks of mobile nationalisms are not benign or purely cultural. They are foundational to the economic and political landscape of both England and the territory left behind.

In geographical writings on landscape, Matless (1998) and Daniels (1993) record the genealogies of cultural landscape values which challenge the 'traditional' elisions between Britishness and English*ness* and England. Within their work, the seemingly benign and objective discourses of Englishness have been reoriented through a critical lens, thus exploding the mythologies of an English landscape culture that is separate from industry, multiracial influences and modernity itself. However, writers such as Ian Chambers (1993) argue that there is more work to be done on the question of race, or more accurately racialized populations that are 'British'. Chambers argues that this is long overdue:

even radical critics and historians such as Raymond Williams, E.P. Thompson, and Eric Hobsbawm have, in their appeals to the continuities of *native* traditions and experiences, perhaps inadvertently conceded the ethical and racial pretensions of a national(ist) mythology. (Chambers, 1993, p. 154)

The British landscape of *Englishness* or *Britishness* has traditionally been situated as a cultural way of being that is 'rooted', 'fixed' and integral to a relationship with land, country, soil and a moral sense of being and understanding this place; writers such as Phil Kinsman (1995), the photographer Ingrid Pollard, and others have attended to the seeming discordancy between Black bodies and the landscape of the English lakes, for example. These bodies are positioned as 'non-native' to this cultural landscape. These writers are challenging the elision between 'native' cultures and 'native natures'. Environmental historians and their 'earth's eye' perspective challenge theories of 'native' and 'non-native' through a consideration and inclusion of both the writing of history through global nature flows and a consideration of nature's own timelines beyond 'nation', 'state' and 'sovereignty'. These environmental histories have centred nature and 'others' that are no longer on the margins of history. By taking this approach to the complexities of the material landscape, we can encounter a notion of geographical landscape history where both culture, nature and material landscape evidence the multicultural nature of British nature and that the cultural record is revisited with a new lens that recognizes the manifestations of material realities of the presence of post-colonial peoples, natures and landscape values.

The culturally determined categories of 'race' continue to haunt the notion of authentic belonging and citizenship. However, these discourses and definitions of 'native' and 'non-native' have various manifestations and genealogies themselves. Ecologists, botanists and landscape historians have all contributed to the debates about how 'non-native' species are invading, exhausting resources, and utilizing space traditionally occupied by 'native' plants and thus native cultural landscapes. These arguments are familiar to us because they have all been exercised in the cultural sphere, especially in the context of discourses on race exemplified in Britain, in the work of Stuart Hall (1990, 1999) and Paul Gilroy (1987, 1993, 2000). The roots of these discourses are in the practices of imperial governance and the cultural logic of colonial definitions of aboriginal peoples of every continent. It has been consistently shown in historical and cultural investigations that in this sphere the questions of 'what is native?' and 'what is not?' are culturally defined. In ecological terms, arguments are raging about which plants contribute to a notion of 'native' and 'non-native' species. Central to these definitions are questions of time. At which point do circulatory and dynamically mobile people or plant species *become* 'native' or natural to our landscape?

Writers such as Nash and Gilroy describe this as the re-emergence of the biological essentialism that is embedded in genetic science and feeds cultural racism. Gilroy (2000) termed this the age of the 'rebirth' of biologism (p. 34), which fosters biologically located evidence of race difference and notions of the purity of genetic stock. These moves ultimately serve to bolster political discourses that seek to define and reassert notions of 'native values' and thus conflate issues of 'cultural belonging'. This sense of belonging is differently played out for racialized communities living in Britain. Belonging is hierarchical. Agyeman (1990, 1991) has long criticized the language of native and non-native as representing a form of ecological racism, arguing that there is a process of 'containment' in operation, keeping the ethnic racialized populations in specified landscapes, namely urban, and usually the most environmentally hazardous. These cultural narratives are contrary to our embracing of notions of 'cosmopolitanism' that are celebrated alongside the benefits of globalization. There is a discordancy between our acceptance of the mobility of bodies across space and the narrating of national belonging and cultures of citizenship that incorporate mobile values and mobile British citizens. Nigel Clark (2002), reflecting on the new 'cosmopolitanism, has recently argued that there is a segregation between 'nature' and 'culture' in these new theorizations. What is important here is that social and cultural dimensions of the global process cannot be severed from the 'non-human'. Ultimately, we need to recognise that 'disturbance, like mobility, invasion and hybridization is endemic to the living world and thus, if it is the nature of "life" to stick to its own "turf" then why are so many species from the taxonomic spectrum so eager for relocation and so well disposed to it?' (p. 103). However, only a partial account of 'citizen', 'history' and 'culture' are present in our public perception, discourse and national institutions (museums, literature and galleries).

Throughout the research, I have attempted to demonstrate the variant formations of Englishness, expressed through diasporic cultures of landscape, as they are embedded within the cultures of everyday life. Through these, English*ness* is in continual process of being remade and reformulated. These new formations are linked dialectically to past modes of engagement with cultures of English landscape both abroad and now at *home*. This is not a new mode for the culture of Englishness, which is memorialized (and has in turn been situated throughout the colonial territories and rememorialized) through experiences in the physical landscapes of England. Mobile bodies, imaginaries and cultures of landscape have historically shifted and shaped cultures of landscape through which we recognize *Englishness,* yet these are never unitary or exhaustive.

> [t]here is not one absolute landscape here, but a series of related, contradictory moments – perspectives – which cohere in what can be recognised as landscape as a cultural process. (Hirsch and O'Hanlon, 1995, p. 23)

The aesthetics, histories and cultures of British Asian-English landscapes stretch out to Lake Nevasha, the Indian Ocean and the English lakes, and are locked into a circulatory process of transformation. Thus, English*ness* is presented here as historically mobile, yet made tangible through cultural materials that have retained currency through a cultural politics of a particular set of landscape aesthetics. A search for a rooted and singular English landscape iconography is the stuff of 'pre-mobile' senses of place; a nonsensical, immoral culture of *an* English nature, body and landscape, belied by the variant landscapes of British Asian Englishness.

What is required in this post-colonial era of citizenry is a geographical understanding of identity politics. Not geographical citizenship as stated in treaties and governance but an understanding of an *environmental citizenship*. Citizenship for British Asians is where 'belonging', attachment, dwelling and heritage are co-constructed through the encounter with place; that is, with the aesthetics, form and grammars of nature; with landscape; and with the process of connecting with them through memory and body. This environmental citizenship is expressed in this research as an everyday lived notion of social heritage, where 'memory' is the formation of social history, and where landscape connotes citizenship to a *territory of culture.*

Bibliography

Agyeman, J. (1990), 'Black People in a White Landscape: Social and Environmental Justice', *Built Environment* 16: 3, 232–36.

Agyeman, J. (1991), 'The Multicultural City Ecosystem, Streetwise', *Magazine of Urban Studies* 7 (Summer), 21–24.

Chambers, I. (1993), 'Narratives of Nationalism: Being "British"', in E. Carter, J. Donald and J. Squires (eds), *Space and Place: Theories of Identity and Location* (London: Lawrence and Wishart), p. 154.

Clark, N. (2002), 'The Demon-Seed: Bioinvasion as the Unsettling of Environmental Cosmopolitanism', *Theory, Culture and Society* 19: 1–2, 101–25.

Crosby, A. (2004), *Ecological Imperialism* (Cambridge: Cambridge University Press).

Daniels, S. (1993), *Fields of Vision: Landscape Imagery and National Identity in England and the United States* (Cambridge: Polity Press).

Daniels, S. (1989), *Fields of Vision* (London: Polity Press).

Gilroy, P. (2000), *Against Race* (Cambridge, MA: Harvard University Press).

Gilroy, P. (1993), *The Black Atlantic: Modernity and Double Consciousness* (London: Verso).

Gilroy, P. (1991), '"It Ain't Where You're from, It's Where You're At ...": The Dialectics of Diasporic Identification,' *Third Text* 13 (Winter), 3–16.

Gilroy, P. (1990), 'Art of Darkness: Black Art and the Problem of Belonging to England', *Third Text* 10 (Spring), 42–52.

Gilroy, P. (1987), *'There Ain't No Black in the Union Jack': The Cultural Politics of Race and Nation* (London: Routledge).

Hall, S. (1999), 'Whose Heritage?: Unsettling "The Heritage", Re-imagining the Post-nation', *Third Text* 49 (Winter), 3–13.

Hall, S. (1990), 'Cultural Identity and Diaspora', in J. Rutherford (ed.), *Identity: Community, Culture, Difference* (London: Lawrence and Wishart), pp. 222–37.

Hall, S. and DuGay, P. (eds) (1996), *Questions of Cultural Identity* (London: Sage).

Hirsch, E. and O'Hanlon, M. (eds) (1995), *The Anthropology of Landscape: Perspectives on Place and Space.* (Oxford: Clarendon Press).

Kinsman, P. (1995), 'Landscape, Race and National Identity: The Photography of Ingrid Pollard', *Area* 27, 300–10.

Massey, D. (2006), 'Landscape as Provocation: Reflections on Moving Mountains', *Journal of Material Culture* 11, 1–2, 33–48.

Matless, D. (1998), *Landscape and Englishness* (London: Reaktion).

Nash, C. (2005), 'Geographies of Relatedness', *Transactions of the Institute of British Geographers* 30: 4, 449–62.

Ritvo, H. (1992), 'Race, Breeds and Myths of Origin', *Representations* 39 (Summer), 1–22.

Appendices

**Paintings and descriptions from the visual workshops conducted
with the British Asian women and landscape artist Melanie Carvalho**

Describe below, in writing, or a drawing, or however you
like, the landscape that best represents your idea of
home.

Please return to: Flat 7, 51 Tudor Road, London E9 7SN

*Pakistan – Lahore. My childhood was in this house, with
huge garden. My dream is to my children play in this
huge garden. Children playing with swimming pool.
Smell: Chameli – white flower. Trees – Jasmine, Vine.
Weather – Neither very hot or very cold. Grape fruit – tangerine*

Appendix 1   Zubeida's description

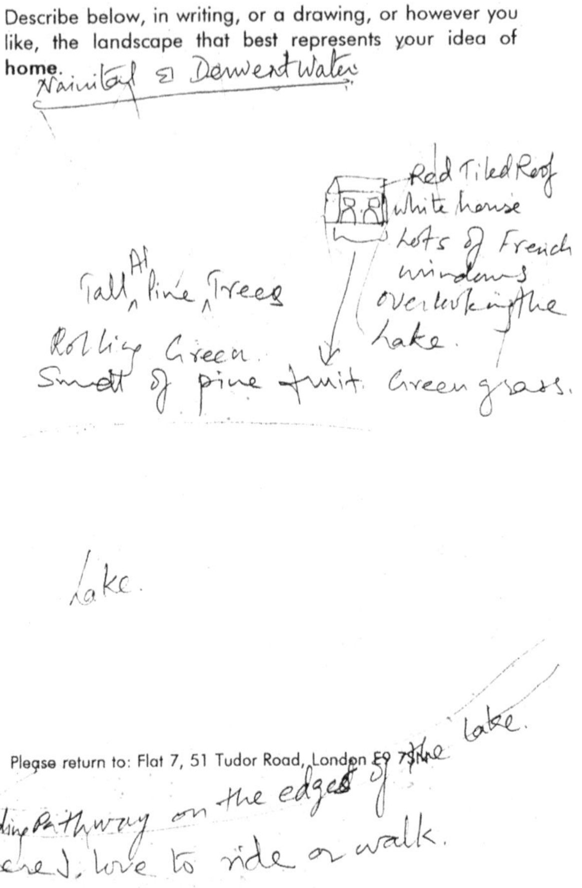

Describe below, in writing, or a drawing, or however you like, the landscape that best represents your idea of home. Nainital & Derwent Water

Red Tiled Roof
White house
Lots of French
windows
overlooking the
Lake.

Tall Alpine Trees

Rolling Green.
Smell of pine fruit. Green grass.

Lake.

the lake.

Winding Pathway on the edges of
which I love to ride or walk.

Appendix 2   Lalita's description

Appendix 3   Carvalho's painting of Lalita's description

Describe below, in writing, or a drawing, or however you like, the landscape that best represents your idea of home.

Please return to: Flat 7, 51 Tudor Road, London E9 7SN

Appendix 4 Puja's description

Appendix 5   Carvalho's painting of Puja's description

Describe below, in writing, or a drawing, or however you
like, the landscape that best represents your idea of
home.

My thoughts of Uganda — Memories of our Tea estates
where I spent most of my school holidays — Beautiful
round big house on top of a hill — About 60 steps
to go to the house — Two muddy roads leading to
the house — Big Badminton Court on side where
cars were also parked — Cement benches around the
Court. Hills — fully green well grassed — from the
house could see our tea factory about half mile
down — smoke from chimney of factory — rough
drawing.

Please return to: Flat 7, 51 Tudor Road, London E9 7SN

You could see estate from back of the house
about mile away — and jungles around.
Just would like to go and spend some there
again if possible.

Appendix 6   Manjula's description

Appendix 7   Carvalho's painting of Manjula's description

Describe below, in writing, or a drawing, or however you like, the landscape that best represents your idea of home.

Please return to: Flat 7, 51 Tudor Road, London E9 7SN

Appendix 8 Darshna's description

Appendix 9 Carvalho's painting of Darshna's description

like, the landscape that best represents your idea of home.

We friends + families go for Picnic to the place is called Thika Falls + Fourteen Falls. There are small Falls water streaming down from big Large stones in different places and people go under the falls to bath. Children also join. Underneath are big smooth stones on which people sit get wet children play under the water streams. We go there for fun. eat, Sing + dance. This is my childhood image. Used to go there at Least twice a year in group consisting of 40 to 50 people young & old.

Big place, big three, 14 falls of water real Thika (falls) undereath smooth rocks/stones, picnic place, safe.

9/10/11/12/5d.
Go in one truck people sit at rear back also a the car for other people.
Gramaphone records with big.

records - 50's films

Very big trees with plenty of leaves for people to sit in the shade.

MUCKO - mat (vash) for we eat ... he and trees.

Nobody are wearing costumes, fully clothed.
food - plenty of food

Appendix 10   Shanta's description

Appendix 11 Carvalho's painting of Shanta's description

Appendix 12   Carvalho's painting of Hansa's description

Appendix 13   Carvalho's painting of Anila's description

Appendix 14   Carvalho's painting of Kajal's description

Appendix 15   Carvalho's painting of Sarabjit's description

Index